Konrad Staudt

Comparative investigations of the sandfish's epidermis

Konrad Staudt

Comparative investigations of the sandfish's epidermis

Surface and molecular examinations of the epidermis of the sandfish (Squamata: Scincidae: Scincus scincus)

Südwestdeutscher Verlag für Hochschulschriften

Impressum / Imprint

Bibliografische Information der Deutschen Nationalbibliothek: Die Deutsche Nationalbibliothek verzeichnet diese Publikation in der Deutschen Nationalbibliografie; detaillierte bibliografische Daten sind im Internet über http://dnb.d-nb.de abrufbar.

Alle in diesem Buch genannten Marken und Produktnamen unterliegen warenzeichen-, marken- oder patentrechtlichem Schutz bzw. sind Warenzeichen oder eingetragene Warenzeichen der jeweiligen Inhaber. Die Wiedergabe von Marken, Produktnamen, Gebrauchsnamen, Handelsnamen, Warenbezeichnungen u.s.w. in diesem Werk berechtigt auch ohne besondere Kennzeichnung nicht zu der Annahme, dass solche Namen im Sinne der Warenzeichen- und Markenschutzgesetzgebung als frei zu betrachten wären und daher von jedermann benutzt werden dürften.

Bibliographic information published by the Deutsche Nationalbibliothek: The Deutsche Nationalbibliothek lists this publication in the Deutsche Nationalbibliografie; detailed bibliographic data are available in the Internet at http://dnb.d-nb.de.

Any brand names and product names mentioned in this book are subject to trademark, brand or patent protection and are trademarks or registered trademarks of their respective holders. The use of brand names, product names, common names, trade names, product descriptions etc. even without a particular marking in this works is in no way to be construed to mean that such names may be regarded as unrestricted in respect of trademark and brand protection legislation and could thus be used by anyone.

Coverbild / Cover image: www.ingimage.com

Verlag / Publisher:
Südwestdeutscher Verlag für Hochschulschriften
ist ein Imprint der / is a trademark of
AV Akademikerverlag GmbH & Co. KG
Heinrich-Böcking-Str. 6-8, 66121 Saarbrücken, Deutschland / Germany
Email: info@svh-verlag.de

Herstellung: siehe letzte Seite /
Printed at: see last page
ISBN: 978-3-8381-3226-6

Copyright © 2012 AV Akademikerverlag GmbH & Co. KG
Alle Rechte vorbehalten. / All rights reserved. Saarbrücken 2012

Content

Abstract		*9*
1.	*Introduction*	*11*
1.1	Comparative taxonomy of *Scincus scincus*	11
1.2	Skin properties of *Scincus scincus*	12
1.4	The epidermis of *Scincus scincus*	14
1.3	Biomimetic aim	17
1.5	Aims of the thesis	18
2.	*Materials and Methods*	*19*
2.1	Materials	19
2.1.1	Chemicals and enzymes	19
2.1.2	Devices	22
2.1.3	Kits	23
2.1.4	Buffers and solutions	23
2.1.5	Primers	27
2.1.6	Animals and skin samples	28
2.2	Surface investigations	28
2.2.1	Comparative friction angle measurements	28
2.2.2	Scanning electron microscopy	29
2.2.3	Atomic force microscopy	30
2.2.4	Skin replicas	30
2.2.5	Abrasion resistance	30
2.2.6	Other possible functions of dorsal serrations	31
2.3	DNA analysis	31
2.3.1	General methods	32
2.3.2	Rapid amplification of cDNA ends (RACE)	33
2.3.3	RNA hybridization	36
2.3.4	T4 RNA ligation	39
2.3.5	Amplification with semi-specific primers	41
2.3.6	Sequencing of a β-keratin coding gene fragment of *Meroles anchietae*	45
2.4	Proteomics	45
2.4.1	General methods	45
2.4.2	Visualization of β-keratin proteins through specific antibodies	47
2.4.3	Visualization of glycoproteins	47
2.4.4	Comparative glycosylation intensities	47
2.4.5	Enzymatical deglycosylation	48
2.4.6	Chemical deglycosylation through β-elimination	49
2.5	Glycomics	51
2.5.1	Prediction of glycosylation sites	51
2.5.2	Silanisation	51

2.5.3	Surface characteristics of immobilised glycans	52
2.5.4	Quantification of linked glycans	53
2.5.5	Glycan analysis	53
2.5.6	Monosaccharide classification	55
2.5.7	Glycopeptide analysis	55
3.	***Results***	***56***
3.1	Surface investigations	56
3.1.1	Comparative friction angle measurements	56
3.1.2	Topography investigation by scanning electron microscopy	57
3.1.3	Surface properties of resin replicas of *Scincus scincus*	57
3.1.4	Other possible functions of dorsal serrations	58
3.2	DNA analysis	60
3.2.1	Rapid amplification of cDNA ends (RACE)	60
3.2.2	RNA hybridization	60
3.2.3	T4 RNA ligation	61
3.2.4	Amplification with semi-specific primers	61
3.2.5	Sequencing of a β-keratin coding gene fragment of *Meroles anchietae*	62
3.3	Proteomics	63
3.3.1	In-silico translation of β-keratin coding DNA sequences	63
3.3.2	Visualization of β-keratin proteins through specific antibodies	65
3.3.3	Comparative glycosylation intensities	65
3.3.4	Enzymatical deglycosylation	68
3.3.5	Chemical deglycosylation through β-elimination	69
3.4	Glycomics	70
3.4.1	Prediction of glycosylation sites	70
3.4.2	Surface characteristics of adsorbed proteins and silanised glycans on glass	71
3.4.3	Quantification of silanised glycans	75
3.4.4	Monosaccharide classification	76
3.4.5	Glycan analysis	76
3.4.6	Glycopeptide analysis	77
4.	***Discussion***	***78***
4.1	Adaptations of *Scincus scincus* to a subterranean life	78
4.2	Microornamentaion of sandfish scales as cause for friction reduction	79
4.3	Possible functions of the microstructure on dorsal scales	79
4.4	Chemical composition of the scales as cause of friction reduction	80
4.5	Analysis of the scale material	80
4.5.1	Biochemichal investigations	80
4.5.2	DNA analysis	81
4.5.3	Glycan analysis	83
4.6	Glycans as cause for friction reduction	84
4.7	Outlook	85

References *87*
Acknowledgements *93*
Bibliographic notes *94*
Appendix *95*

List of figures

1.1	Taxonomy of *Scincus scincus* 12
1.2	Comparison of friction angles between *Scincus scincus* and technical surfaces 13
1.3	AFM measurements of *S. scincus* (ventral and dorsal) and a corn snake 14
1.4	Microstructure of different sandswimming reptiles 15
1.5	Cross section of the outer lepidosauran scale 15
1.6	PAS and Coomassie-blue stain of an exuviae lysates 16
2.1	See-saw for friction angle measurements 29
2.2	Schematic method of RACE 34
2.3	Schematic method of RNA hybridization 37
2.4	Schematic method of T4 RNA ligation 40
2.5	Amplification of genomic DNA with gene specific primers 42
2.6	Sequence map of two aligned β-keratin gene parts 43
2.7	Method to prove that amplified DNA contains β-keratin coding genes 44
2.8	Chemical deglycosylation through β-elimination 50
2.9	Silanisation of glycans on glass via APTES 52
3.1	Friction angles of different reptiles 56
3.2	SEM images of ventral and dorsal scales 57
3.3	Resin replicas of different scale locations 58
3.4	Abrasion resistance of replicas 58
3.5	Transmission in the electromagnetic spectrum of scales 59
3.6	Transcribed RNA by T7 RNA polymerase 60
3.7	Sequence after RNA hybridization 61
3.8	Sequence of a complete β-keratin gene of *S. scincus* and *E. scneideri* 62
3.9	Sequence of a β-keratin gene fragment of *M. anchietae* 63
3.10	β-keratin alignment of *P. siculus*, *S. scincus* and *E. schneideri* 63
3.11	β-keratin alignment of *S. scincus* and *E. schneideri* 64
3.12	β-keratin alignment of *M. anchietae* and *P. siculus* 64
3.13	Reactions with β-keratin specific antibodies 65
3.14	ECL visualization at 5 µg protein quantity 66
3.15	Evaluation of lectin staining intensities 67
3.16	ECL visualization at 10 µg protein quantity 68
3.17	Enzymatical deglycosylation 69
3.18	Control of glycan purification 70
3.19	O-glycosylation potential 71

3.20	Comparative adhesion force of glycans and proteins	72
3.21	Adhesion force vs. friction angle	73
3.22	Abrasion resistance of silanised glass coverslips	74
3.23	Quantification of linked carbohydrates on glass	75

List of tables

1.1	Lectin reactions with β-keratins of *S. scincus*	17
3.2	Monosaccharide analysis of exuvial proteins of *S. scincus*	76

Abstract

The sandfish (Scincidae: *Scincus scincus*) is a lizard capable of moving through desert sand in a swimming-like fashion. The epidermis of this lizard shows a low friction to sand as an adaption to a subterranean life below the desert's surface. Caused by material properties of β-keratin proteins forming the outer epidermis, this low friction reduces further adhesive wear. Both skin effects, the friction reduction and abrasion resistance outperform even steel. A possible explanation for these properties is an increased glycosylation of the β-keratins. In this study, the friction and the micro-structure of the epidermis as well as the β-keratin coding DNA and the glycosylation of the β- keratin proteins of the sandfish was investigated in comparison to other sauropsidean species, mainly with the closely related Berber skink (Scincidae: *Eumeces schneideri*) and another sandswimming species, the not closer related Shovel-snouted lizard (Lacertidae: *Meroles anchietae*). The friction angle to sand of the examined skins of sandswimming species was found to be much lower in comparison with other reptiles. The micro-structure investigated was very similar between the sandfish and closely related non-sandswimming skinks, and no specific micro-structure was identifiable in the sandswimmer *M. anchietae*; giving further evidence that abrasion resistance and low friction is not caused by the micro-structure of the epidermis, but by material properties of the β-keratins. Furthermore, the glycosylation of β-keratins in various sauropsidean species was investigated and all species tested were found positive; however, the sandswimming species examined show a much stronger glycosylation of their β-keratins. In order to prove this finding through a genetic foundation, the DNA of a complete β-keratin coding gene of the *S. scincus* and a fragment of *M. anchietae* was sequenced and compared with homologue genes of the related species *E. schneideri* and *Podarcis siculus* (Lacertidae). By comparison of the in-silico translated protein sequence, a higher abundance of O-glycosylation sites was found (enabled through the amino acids serine and threonine), giving molecular support for a higher glycosylation of the β-keratin in both sandswimming species. Glycan based friction reduction could also be verified by force-distance measurements via atomic force microscopy. For this, (1) a keratinolysis of sandfish exuviae was performed and the proteins reconstituated on a glass surface and (2) proteins were deglycosylated and the glycans covalently bound on a glass surface by silanisation. Both surfaces showed low adhesion force similar to the untreated skin of the sandfish in comparison to *Eumeces schenideri* and a glass control. Furthermore, a positive impact on abrasion resistance could be observed. Presumably, the glycan structure reduces adhesion forces between the outer epidermis and sand through prevention of the formation of van der Waals bonds. A monosaccharide analysis revealed carbohydrates that are typically present in glycans.

Abstract

Der Sandfisch (Scincidae: *Scincus scincus*) ist eine Echse der es möglich ist, sich in einer schwimmähnlichen Weise durch Sand zu bewegen. Die Epidermis dieses Skinks zeigt einen sehr niedrigen Reibungswinkel zu Sand als eine Adaptation seines Lebens unter der Wüstenoberfläche. Die niedrige Reibung, welche durch Eigenschaften des β-keratins verursacht wird, reduziert weiterhin den Verschleiß der Haut. Beide Eigenschaften der Epidermis, die niedrige Reibung und die hohe Verschleißresistenz übertreffen sogar die Werte von Stahl. Eine mögliche Erklärung für diese Effekte ist eine Glykosylierung der β-keratine. In diesem Promotionsprojekt wurden die Reibung und die Mikrostrukturierung der Epidermis, als auch die β-keratin kodierende DNA und die Glykosilierung der β-keratin Proteine des Sandfisches untersucht. Verglichen wurden die Ergebnisse mit weiteren Vertretern der Sauropsidia, jedoch hauptsächlich mit dem nahe verwandten Berberskink (*Eumeces schneideri*) und einem weiteren, konvergent evolvierten Sandschwimmer, der Dünenechse (Lacertidae: *Meroles anchietae*). Die Reibung der Haut zu Sand der untersuchten sandschwimmenden Reptilien war sehr viel niedriger im Vergleich zu anderen Reptilien, wobei die Mikrostruktur sehr ähnlich bei den untersuchten Skinken war; es konnte jedoch keine Strukturierung bei *Meroles anchietae* nachgewiesen werden. Diese Ergebnisse gaben einen weiteren Beweis dafür, dass nicht die Oberflächenstruktur, sondern die chemische Beschaffenheit der Epidermis als Ursache für die niedrige Reibung und die Abreibungsresistenz der Haut des Sandfisches in Frage kommt. Daraufhin wurde die Glykosilierung der β-keratine verschiedener Sauropsiden (Schildkröten, Vögel, Schlangen, Echsen) quantitativ erfasst. Es stellte sich heraus, dass die β-keratine aller Reptilien eine Glykosylierung aufweisen, jedoch diese bei den Sandschwimmern weitaus stärker ausgeprägt ist. Um dieses Ergebnis genetisch zu überprüfen, wurde die β-keratin kodierende DNA des Sandfisches und des nah verwandten, nicht-sandschwimmenden *Eumeces schneideri* sequenziert und verglichen. Weiterhin konnte ein Fragment eines β-keratin Gens von *Meroles anchietae* mit dem Gen des nah verwandten *Podarcis siculus* verglichen werden. Es konnte nachgewiesen werden, dass im Vergleich zu den nicht-sandschwimmenden Arten eine höhere Abundanz der O-glykosylierungsstellen (ermöglicht durch die Aminosäuren Serin und Threonin) in der in-silico übersetzten Proteinsequenz des β-keratins in beiden Sandschwimmern vorhanden ist. Eine Glykan-basierte Reibungsreduktion konnte durch Kraft-Abstandsmessungen via Rasterkraftmikroskopie (AFM) nachgewiesen werden. Hierfür wurde (1) eine Keratolyse der Sandfischhaut durchgeführt und die gelösten Proteine auf einer Glasoberfläche rekonstituiert und (2) die Glykane von dem β-keratin Protein des Sandfisches getrennt, aufgereinigt und anschließend über ein Silan auf eine technische Oberfläche (Glas) kovalent gebunden. Beide erzeugten Oberflächen zeigten dieselbe niedrige Adhäsionskraft wie die native Haut und es konnte eine deutliche Erhöhung der Abreibungsresistenz des mit Glykanen beschichteten Glases festgestellt werden. Durch die Kraft-Abstandsmessungen konnte gezeigt werden, dass die niedrige Reibung in Zusammenhang mit einer niedrigen Adhäsionskraft steht. Die Ergebnisse lassen darauf schliessen dass die Glykane die Adhäsionskräfte zwischen der Haut und anderen Materialien vermindern; möglicherweise geschieht dies durch die Verhinderung der Formierung von van der Waals Dipolen. Eine Analyse der Monosaccharide resultierte in Kohlenhydraten welche typischerweise Glykane zusammensetzen.

1. Introduction

World's deserts are a hostile habitat due to the absence of water and high temperatures. Animals that are able to survive in this extreme environment often evolved remarkable adaptations to avoid heat and predation. Species like the sandfish (Scincidae: *Scincus scincus*) adapted the fascinating feature of a subterranean life. While many desert-living reptile species have the ability to bury themselves into sand (i.e., some species of the genus Phrynosoma), these skinks spend most of the time below the hot desert surface in colder sand layers and are able to move with a high velocity up to 30 cm/s over considerable distances through aeolian sand (Baumgartner et al., 2008; Maladen et al., 2009). Reptilian locomotion is a complex system including many factors, like direction of movement, anisotropy, body form and skin characteristics (i.e., Hazel et al., 1999; Berthe et al., 2009). In case of the sandfish's locomotion, it could be shown that with its body undulations of 3 Hz a decompaction of sand will emerge and this medium will behave more like a fluid than a solid body (Baumgartner et al., 2008), helping the sandfish to move through sand. This amazing ability, which is called "sandswimming", is facilitated through morphological adaptations like a shovel shaped snout, strong limbs with fringed digits and protected nostrils (Hartmann, 1989). Furthermore, the skin shows a high resistance against abrasion together with a low friction angle to sand in comparison to other materials (Rechenberg and El Khyari, 2004; Baumgartner et al., 2007). These properties of the skin are of great biological, chemical and industrial interest and the investigation of the physical-chemical causes of low friction and high abrasion resistance the main interest of this thesis. After a short introduction into the biology of the sandfish, the aim of this thesis is to compare the surface and molecular properties of the skin with other sauropsidean species, mainly with the closely related Berber skink (Scincidae: *Eumeces schneideri*) and the not closer related Shovel-snouted lizard (Lacertidae: *Meroles anchietae*). This is approached by investigations of the skins' friction angles and ultra-structures, β-keratin coding DNA, proteomics of the β-keratins and glycomics of the glycans linked to the β-keratins.

1.1 Comparative taxonomy of *Scincus scincus*

The family Scincidae is with the Gekkonidae one of the most diverse lizard group with more than 1300 species (Bauer, 1998). *Scincus* is a genus in the family Scincidae inside the order squamata. Next to the sandfish *Scincus scincus*, this genus has two more species, *Scincus mitranus* and *Scincus hemprichii* (Arnold et al., 1977). The taxonomic status of *Scincus albifasciatus* is unsettled (Kalboussi et al., 2006); though gene flow between *S. scincus* and *S.*

Introduction

albifasciatus may occur, both are regarded as different species in literature. The sandfish inhabits an arid environment with aeolian sand and is found from Northern Africa (Sahara desert) to the east into Saudi Arabia, Iran and Iraq (Kohlmeyer, 2001). It nourishes mainly on small arthropods and plant seeds (Al-Sadoon et al., 1999). Genetic investigations revealed that *Eumeces schneideri*, *Eumeces algeriensis* and *Scincopus fasciatus* are closest related to the genus *Scincus* (Schmitz et al., 2004) (fig. 1.1). The closely related Berber skink (*Eumeces schneideri*) is found in stony habitats from Northern Africa to South-west Asia. Though this species is able to bury itself into lose ground material, this species does not show morphological adaptations to a subterranean life like the sandfish. The close relationship between both species enables the Berber skink for comparative studies. As third species for comparison used is the Shovel-snouted lizard *Meroles anchietae* (Lacertidae). This species, not closely related to the sandfish, evolved convergently the ability of sandswimming with similar morphological adaptations and lives in the Namib Desert.

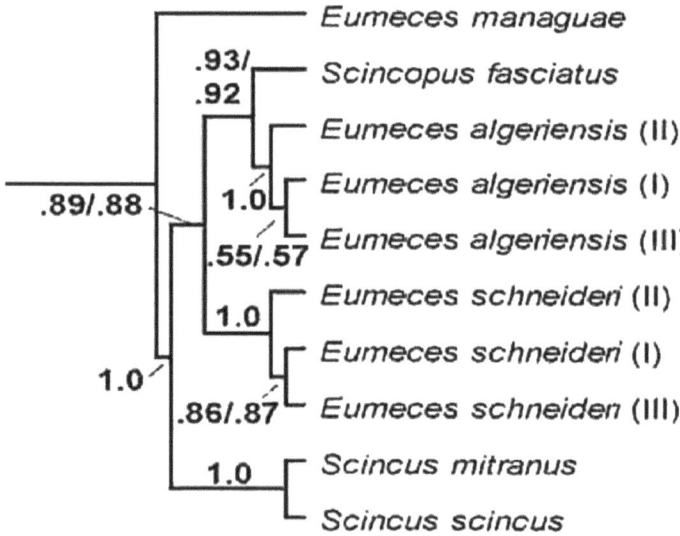

Figure 1.1 – Taxonomic relationship of *Scincus scincus* (Schmitz et al., 2004).

1.2   Skin properties of *Scincus scincus*

The friction angle is defined as the angle at which a granular medium begins to slide off from a surface. Hereby, the sandfish's skin has a low friction angle to sand of only $\theta = 21°$ (corresponding friction coefficient $\mu = \tan \theta = 0.38$), whereas the friction angle of all other

Introduction

materials investigated was found higher: e.g., the friction angle of steel is $\theta = 25°$ ($\mu = 0.47$) and a colubrid epidermis has a friction angle of $\theta = 30°$ ($\mu = 0.58$) (Baumgartner et al. 2007; Rechenberg et al., 2004; Saxe, 2008) (fig.1.2). This low friction allows an energy saving, swimming-like movement through sand.

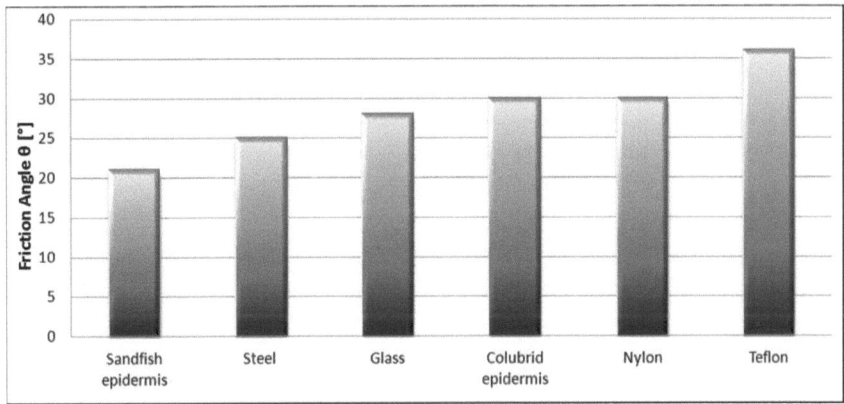

Figure 1.2 – Comparison of the friction angles of technical surfaces with the sandfish and a colubrid epidermis (data after Baumgartner et al., 2007; Rechenberg, et al., 2004; Saxe, 2008).

The low friction angle of the epidermis of the sandfish is correlated with a high resistance against abrasion. The skin was found to be resistant to sandblasting even after the duration of six hours from a height of 25 cm, whereas glass became dull yet after a short time (Rechenberg et al., 2004; Saxe, 2008). Investigation with atomic force microscopy (AFM) resulted in low adhesive force between the epidermis and the tip in force-distance measurements (Baumgartner et al., 2007) (fig. 1.3).

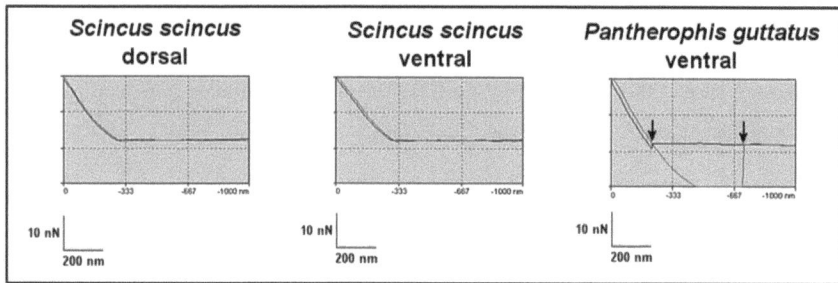

Figure 1.3 – AFM force-distance measurements of the sandfish's epidermis (dorsal and ventral) in comparison to the corn snake (*Pantherophis guttatus*). A clear attractive force can be seen in the snake in retreat movement (marked through arrows) which is missing in the sandfish (Baumgartner, et al., 2007).

1.3 The epidermis of *Scincus scincus*

Dorsal scales show a serrated structure that is suspected to cause the low friction to sand (Rechenberg et al., 2004; Rechenberg et al., 2009). This structure is found not only in the sandfish, but also in other subterranean living reptiles, like the sandboa (*Eryx colubrinus*) or the Wedge-snouted skink (*Sphenops sepsoides*), what raises the question of a convergent evolution towards these micro-serrations on the scales, and whether those are essential for the abilities of abrasion resistance and low friction to sand (Rechenberg et al., 2009) (fig.1.4).

However, scanning electron microscopy (SEM) imaging of several locations on the sandfish skin showed that this structure is only present on dorsal scales and completely absent on ventral scales; nevertheless both locations show the same properties of abrasion resistance and low friction angle. Furthermore, it was demonstrated that low friction and abrasion resistance is a material property by dissolving and reconstituting the keratins on a polymer surface; which resulted in properties similar to the original scale. This is giving evidence that the observed skin properties are independent from surface structures but caused by chemical composition (Baumgartner et al., 2007).

Introduction

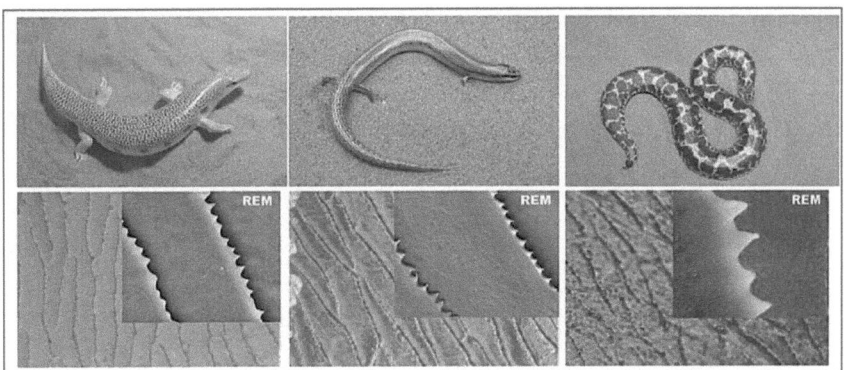

Figure 1.4 – Microstructure of different sandswimming reptiles. The sandfish (*Scincus albifasciatus*, left), the Wedge-snouted skink (*Sphenops sepsoides*, middle) and the sandboa (*Eryx colubrinus*, right) (Rechenberg et al., 2009).

Since it could be shown that material properties are the cause for the skin properties in *S. scincus*, an investigation of the most prominent skin proteins, the β-keratins is necessary. The outer oberhautchen consists mainly of this protein and these proteins are thus in direct contact with the environment (Sawyer et al., 2003) (fig. 1.5). β-keratins are proline-glycine rich that aggregate in a β-sheet conformation and form disulfide bridges (through cystin) in the secondary protein structure, which accounts to the hard corneous outer layer of the epidermis. These proteins have a low molecular weight of 10 to max. 30 kDa (Toni et al., 2007).

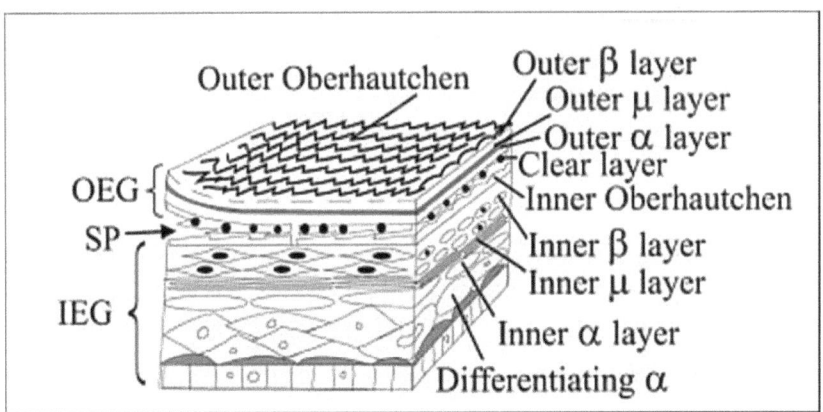

Figure 1.5 – Cross section of the outer scale section in lepidosauran scales (Toni et al., 2007).

Biochemical investigations revealed that β-keratin proteins of the sandfish scales are glycosylated (Baumgartner et al., 2007) (fig.1.6). There are many types of glycoproteins

15

found in any organism. It is estimated that more than 50% of all proteins are glycosylated, which mainly serves for intracellular processes (increase of stability, protection from proteolysis, protein targeting, etc. (i.e., Taylor and Drickamer, 2006)). However, a glycosylation of keratin seems to be unusual and thus might be the key for understanding the characteristics to sand of the sandfish's epidermis.

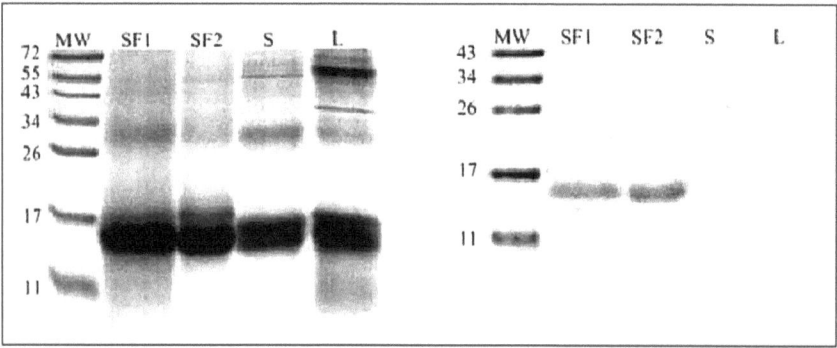

Figure 1.6 – Coomassie-blue (left) and PAS stain (right) of skin lysates of the sandfish (SF1 and SF2), *Pantherophis guttatus* (S) and *Crotapythus collaris* (L). Only the sandfish shows glycosylated β-keratin proteins in the PAS reaction, giving evidence of glycosylation of this protein (Baumgartner et al., 2007).

There are two types of glycosylation possible: N-glycosylated glycans are linked to asparagines in the consensus protein sequence Asn-X-Ser/Thr; O-glycosylated glycans can be linked to serine and threonine (Montreuil, 1982). For this reason, to find out the type and abundance of glycosylation sites within the gene, the DNA sequence with the in-silico translated protein sequence of the β-keratin coding DNA is essential to know. A fragment of the β-keratin gene has been identified in Saxe, 2008 through the use of β-keratin universal primers. Different lectins employed as specific carbohydrate markers can reveal the type of glycosylation (Patsos et al., 2009) and the involved monosaccharides can be visualized in western blot utilizing enhanced chemiluminescense system (ECL). All lectins tested in Saxe, 2008 reacted with β-keratin linked carbohydrates of Scincus scincus (table 1). Every tested lectin can bind on both, N- and O-linked glycans, except PNA that binds on the core 1 disaccharid of O-linked glycans (Galβ3GalNAcα/β) (Goldstein and Hayes, 1987). The positive reaction of MAA and SNA indicate sialic acid attached to terminal galactose in a (α-2,6) or (α-2,3) linkage (Shibuya et al., 1987; Wang and Cummins, 1988). GNA and concanavalin A (conA) are specific for mannose, suggesting a high proportion of this carbohydrate in the glycans (Shibuya et al., 1988). DSA binds specifically to N-acetylglucosamine (Crowley et al., 1984).

Lectin	conA	MAA	SNA	GNA	DSA	PNA
Monosaccharid specificity	Man Glc	Gal	Sia	Man	GlcNAc	Gal
Scincus scincus (β-keratin)	+	+	+	+	+	+
Fetuin (positive control for MAA, SNA, DSA)	-	+	+	-	+	-
BSA (negative control)	-	-	-	-	-	-

Table 1 – Reactions of different lectins with the β-keratin of *S. scincus* (visualized through western blot/ECL system). All lectins tested bound on carbohydrates attached to the β-keratin protein of the sandfish (indicated through a plus in this table), whereas the positive control (fetuin) only reacted with specific lectins (MAA, SNA and DSA) and the negative control (BSA) did not react with a lectin (Saxe, 2008).

1.4 Biomimetic aim

The reduction of wear effects is of great interest in industrial applications. Worldwide, billions of dollars are used to repair erosive and corrosive damages each year (i.e., CC Technologies, 2001), whereby many are caused by tribologic effects. Adhesion is (along with deformation) the main mechanism causing friction (Czichos et al., 1978) and created mainly through van der Waals, capillary and electrostatic forces (Eastman and Da-Ming Zhu, 1996). Abrasion resistance in reptiles was found to be determined by a gradient of harder outer scale layers with a higher elasticity modulus to a softer inner integument (Klein et al., 2010), which may reduce abrasive wear by absorbing kinetic energy caused by locomotion through abrasive surfaces such as sand. However, less friction can reduce adhesive wear as well, which is caused by relative motion and direct contact between two materials (Jones and Scott, 1983).

Being able to mimic the epidermic properties of *S. scincus* on technical surfaces might enhance a massive cost reduction whenever granular medium is moved due to energy decrease in transport and less abrasion. Additionally, constructions that are built in arid environments can be better protected from sand storms, for example solar panel parks in the

Sahara desert. Also, the lifetime of technical components can be increased through coating of the surface with a substance similar to that which is used in the epidermis of *S. scincus*. Moreover, the use of oil as lubricant might become superfluous in many technical applications. Furthermore, it is also possible to protect sensitive surfaces from scratching, which is necessary in i.e., touch screens.

1.5   Aims of the thesis

The aims of this work are to find out more about the comb-like serrated ultra-structure found on dorsal scales of the sandfish. This was performed by comparison of the micro-ornamentation of the sandfish with closely related, but non-sandswimming skinks, the Banded skink (Scincidae: *Scincopus fasciatus*), the Berber Skink (Scincidae: *Eumeces schneideri*) and the not closer related, convergently evolved sandswimming species, the Shovel-snouted lizard (Lacertidae: *Meroles anchietae*). From the sandfish, resin replicas of dorsal scales were produced and tested for abrasion resistance, since resin replicas of a surface often resemble similar physical properties (e.g., Comanns et al. 2011). Finally, other possible function of the comb-like structures, radiation- and moisture harvesting and triboelectricity were tested.

Furthermore, molecular investigations of the β-keratins of the sandfish were performed in comparison to the Berber skink and the Shovel-snouted lizard. The main occupation was to sequence the complete β-keratin coding DNA to find out the abundance and type of glycosylation sites present on the in-silico translated protein. Moreover, preserving the DNA in glycerol stocks enables further biotechnical applications like protein expression in CHO cells to synthesize proteins/glycans similar to those present in the sandfish. Proteomic research was performed by evaluating glycosylation intensities in comparison to other sauropsidean species and by enzymatical and chemical deglycosylation of proteins. Investigations in the field of glycomics were carried out by MALDI-TOF massspectroscopy, monosaccharide analysis and by covalently linking of purified glycans on a technical surface.

2. Materials and Methods

2.1 Materials

2.1.1 Chemicals and enzymes

1 kb gene ruler	Fermentas, St. Leon-Rot, Germany p
2 - Mercaptoethanol, 99%	AppliChem, Darmstadt, Germany
3-Aminopropyltriethoxysilane	Riedel de Haen, Seelze, Germany
6x loading dye	Fermentas, St. Leon-Rot, Germany
Acetic acid, 100 %	Riedel de Haen, Seelze, Germany
Acetone $\geq$ 99.5 %	AppliChem, Darmstadt, Germany
Agar	AppliChem, Darmstadt, Germany
Agarose NEO Ultra Qualität	Roth, Karlsruhe, Germany
Albumine from bovine serum (BSA)	Sigma-Aldrich, München, Germany
Ammonium hydroxide 32%	AppliChem, Darmstadt, Germany
Ammoniumperoxodisulfate (APS)	Roth, Karlsruhe, Germany
Ampicillin (Na-salt), $\geq$99 %	Roth, Karlsruhe, Germany
Borat, $\geq$ 99.8 %, p.a.	Roth, Karlsruhe, Germany
Brillant Blue G250	Roth, Karlsruhe, Germany
Bromphenolblau Na - Salt	Roth, Karlsruhe, Germany
Calciumchloride ($CaCl_2$)	Merk, Darmstadt, Germany
Canning foil	Bringmann, Wendelstein, Germany
Centiprep YM-10	Millipore, Schwalbach, Germany
concanavalin A Peroxidase	Sigma-Aldrich, München, Germany
Cyanoborhydride, $\geq$ 98 %	Fluka, Buchs, Swizerland
Denhardt solution (50 x concentrate)	Serva, Heidelberg, Germany
di – Sodiumhydrogenephosphate-2-hydrat	Merk, Darmstadt, Germany
Dowex 50WX2-400 (H+ form)	Sigma-Aldrich, München, Germany
dCTP Solution (100 mM)	Fermentas St. Leon-Rot, Germany
dGTP Solution (100 mM)	Fermentas St. Leon-Rot, Germany
dNTP Solution (100 mM)	Fermentas St. Leon-Rot, Germany
Diethylene pyrocarbonate (DEPC)	AppliChem, Darmstadt, Germany
Dimethylsulfoxide (DMSO) $\geq$ 99.5%	Sigma-Aldrich, München, Germany
DNase I (from biovine pancreas grade II)	Roche, Mannheim, Germany

Epoxy resin and hardener	Toolcraft Conrad Electronic SE, Hirschau, Germany
Ethanol, ≥ 99.8 %, p.a.	Roth, Karlsruhe, Germany
Fetuin from fetal bovine serum	Sigma-Aldrich, München, Germany
Fish sperm DNA	Serva, Heidelberg, Germany
Formaldehyde (37 % solution)	Merk, Darmstadt, Germany
Formamide 99.5 %	AppliChem, Darmstadt, Germany
Fount India Ink 17, Black	Pelikan, Hannover, Germany
Glycerine, ≥ 86 %, p.a.	Roth, Karlsruhe, Germany
Glycin ≥ 99 %, p.a.	AppliChem, Darmstadt, Germany
Goat anti-rabbit-peroxidase (garb-POX)	Jackson ImmonoResearch, West Grove, PA, USA
Hybond™-c extra nitrocellulose membrane	Amersham (GE healthcare), Freiburg, Germany
Hybond™-c nitrocellulose membrane	Amersham (GE healthcare), Freiburg, Germany
Hybond™-N nylon membrane	Amersham (GE healthcare), Freiburg, Germany
Hyperfilm ECL	Amersham, Buckinghamshire, UK
Hydrogen peroxide, 30% (w/v) in water	Sigma-Aldrich, München, Germany
Isopropanol	Roth, Karlsruhe, Germany
Luminol 98%	Sigma-Aldrich, München, Germany
Magnesiumchloride ($MgCl_2$) ≥ 99%	Roth, Karlsruhe, Germany
Manganchloride ($MnCl_2$)	Merk, Darmstadt, Germany
Methanol	AppliChem, Darmstadt, Germany
Molecular weight marker	Fermentas, St. Leon-Rot, Germany
N-Glycosidase F, recombinant	Roche, Mannheim, Germany
O-Glycosidase	Roche, Mannheim, Germany
Parafilm	American National Can Company
p-Coumaric acid	Sigma-Aldrich, München, Germany
PCR Master mix Solution (i-StarMax II)	Hiss Diagnostics, Freiburg, Germany
PCR-Script Amp Cloning Kit	Stratagene, La Jolla, Canada
Peroxidase-conjugated Streptavidin	Jackson Immuno research labratories,

	Newmarket, Suffolk, UK
Phenylmethylsulfonylfluorid (PMSF)	Roth, Karlsruhe, Germany
Polysorbate 20 (Tween 20)	Roth, Karlsruhe, Germany
Ponceau S	Roth, Karlsruhe, Germany
Potassium dihydrogen phosphate	Roth, Karlsruhe, Germany
Restriction enzymes (Bam HI, EcoRI,)	Fermentas, St. Leon-Rot, Germany
Reverse Transkriptase	Fermentas, St. Leon-Rot, Germany
RiboLock™ RNase Inhibitor	Fermentas, St. Leon-Rot, Germany
Ribonuclease A	Fermentas, St. Leon-Rot, Germany
RNeasy mini kit	Quiagen, Hilden, Germany
Roti Block (10x concentrate)	Roth, Karlsruhe, Germany
Roti Quant (Bradford reagent)	Roth, Karlsruhe, Germany
Rotiphorese Gel 30	Roth, Karlsruhe, Germany
Sambucus nigra lectin (SNA): Biotin	Vector laboratories, Burlingame, Canada
Serva blue G250	Serva, Heidelberg, Germany
Sialidase	Roche, Mannheim, Germany
Skimmed milk	Applichem, Darmstadt, Germany
Sodium chloride, > 99.8 %	Roth, Karlsruhe, Germany
Sodium-cyanoborohydride 1M	Fluka, Buchs, Switzerland
Sodium dihydrogen phosphate dihydrate	Roth, Karlsruhe, Germany
Sodium dodecyl sulfate (SDS), ultra pure, $\geq$ 99 %	Roth, Karlsruhe, Germany
Sodium hydroxide, 1N	Roth, Karlsruhe, Germany
Sulphuric acid, 1M	Sigma-Aldrich, München, Germany
Sybr Safe DNA Gel Stain	Invitrogen, Karlsruhe, Germany
Trans-illumination scanner Aficio 3235C	Ricoh, Hannover, Germany
T4 DNA ligase	Applichem, Darmstadt, Germany
T4 RNA ligase	Fermentas, St. Leon-Rot, Germany
Terminal Transferase	Fermentas, St. Leon-Rot, Germany
Tetramethylbenzidine (TMB ready to Use ELISA substrate)	Serva, Heidelberg, Germany
Tris-(hysroymethiyl)-aminoethan (Tris)	Roth, Karlsruhe, Germany
Tri-sodium-citrate-dihydrate (Na3citrate)	Roth, Karlsruhe, Germany

Tryptone	AppliChem, Darmstadt, Germany
Tween 20	Roth, Karlsruhe, Germany
Urea (Harnstoff)	Sigma-Aldrich, München, Germany
VPS Hydro	Henry Schein INC. Melville USA
X-Ray developer	Kodak, Stuttgart, Germany
X-Ray Fixer	Kodak, Stuttgart, Germany

2.1.2 Devices

Atomic force microscopy system "Explorer"	Veeco Instruments, Woodbury, NY, USA
Biophotometer plus (DNA concentration)	Eppendorf, Hamburg, Germany
Heating block	Stuart Scientific, Stone, UK
Heating plate TW27	Hartenstein, Würzburg, Germany
Horicontal shaker 3017 GFC	Burgwedel, Germany
Magnetic stirrer	IKA-Werke, Staufen, Germany
MyCycler (PCR)	Bio-Rad, München, Germany
i-scan digitizer	ISS Group Services Ltd., Manchester, UK
PAGE aparatus Mini Protean II	Bio-Rad, München, Germany
Photometer PCP 6121	Eppendorf, Hamburg, Germany
Power source PPC 300/200.4	Northumbria Biologicals Limited, Cramlington, UK
Spektralphotometer U-2000	Hitachi, Tokyo, Japan
Scanning electron microscope Stereoscan 604	Cambridge Instruments, Cambridge, UK
Semidry blot apparatus Trans-Blot SD Cell	Bio-Rad, München, Germany
Sputter coater	Hummer Technics Inc., Alexandria, USA
Table top centrifuge 5415C	Eppendorf, Hamburg, Germany
Thermal controler, enviromental shaker ES 20	MS Laborgeräte, Wiesloch, Germany
Thermocycler	MWG Biotech, Ebersberg, Germany
Trans-illumination scanner Aficio 3235C	Ricoh, Hannover, Germany
UV table TFP-M/WL	LTF Labortechnik, Wasserburg,

	Germany
Vacufuge Concentrator 5301	Eppendorf, Hamburg, Germany
X-Ray cassette 24x30 cm	Dr. Goos-Suprema, Heidelberg, Germany

2.1.3 Kits

High Pure PCR Product Purification Kit	Roche, Mannheim, Germany
High Pure Plasmid Isolation Kit	Roche, Mannheim, Germany
Geno/mini DNA Isolation Spin Kit	Applichem, Darmstadt, Germany
T7-Transkription Kit	Fermentas, St. Leon-Rot, Germany
RNeasy mini Kit	Quiagen, Hilden, Germany

2.1.4 Buffers and solutions

Coomassie-blue staining solution
- 2.5 g Serva Blue G250
- 454 ml isopropanol
- 92 ml acetic acid
- Ad 1.0 l dH$_2$O

Coomassie destaining solution
- 50 ml acetic acid
- 75 ml ethanol
- Ad 1.0 l dH$_2$O

Deglycosylation buffer
- 75 µl 1 M NaHPO4
- 15 µl 0.5 M EDTA
- 7.5 µl Triton X
- 7.5 µl 10% SDS
- 7.5 µl 2-mercaptoethanol
- 5 µl dH$_2$O
- 32,5 µl 9 M Urea

Diethylpyrocarbonate (DEPC)-water
- 0.1 % (v/v) DEPC in 1.0 l dH_2O was incubated over night with lid not completely closed and autoclaved the next day

Electrophoresis buffer for polyacrylamide gels
- 5 g SDS
- 75 g glycin
- 15 g Tris
- Ad 1.0 l dH_2O

Enhanced chemiluminescence (ECL) solution I
- 1 ml 250 mM Luminol (3-Aminophthalhydrazide) in DMSO
- 0.44 ml 90 mM p-Coumaric acid (4-Hydrocinnamic acid) in DMSO
- 10 ml 1 M Tris/HCl, pH 8.5
- Ad 100 ml dH_2O

ECL solution II
- 64 µl 30 % H_2O_2
- 10 ml 1 M Tris/HCl, pH 8.5
- Ad 10 ml dH_2O

Glycerol-stocks from bacterial culture
250 µl of 80 % glycerol in LB-medium are added to 750 µl of an overnight culture and immediately frozen and stored at -80 °C.

Ink staining solution (Holtzauer, 1997)
- 100 ml PBS
- 0.25 g Tween 20
- 0.5 ml Indian ink

Laemmli buffer 3x with 2.5% 2-mercaptoethanol (Laemmli, 1970)
- 0.454 g Tris
- 1.2 g SDS
- 7.56 g glycin

- 400 µl bromphenol blue (2 mg / ml)
- 500 µl 2-mercaptoethanol
- Ad 20 ml dH$_2$O

Lysis buffer (Alibardi et al., 2006)
- 267 µl 9M Urea
- 20 µl 1.5M Tris/HCl pH 7.6
- 5.2 µl PMSF
- 2.1µl 2-mercaptoethanol
- 5.7 µl dH$_2$O

LB-Agar
- Autoclave 15 g/l agar in LB-medium

Lysogeny broth (LB)-medium, pH 7.5
- 10 g BactoTryptone
- 10 g NaCl
- 5 g Yeast extract
- Ad 1 l dH$_2$O
- pH was adjusted with 1 N NaOH and medium was autoclaved on the same day.

MOPS-buffer 10 x
- 0.2 M 3[morpholino]-propane-sulfonic acid
- 0.05 M Natriumacetate, pH 7.0
- 0.01 M Na$_2$EDTA

Phosphate buffered saline (PBS), pH 7.3
- 40.03 g NaCl
- 1.0 g KCl
- 8.5 g Na$_2$HPO$_4$
- 1.0 g KH$_2$PO$_4$
- Ad 5.0 l dH$_2$O

Ponceau S red staining solution
- 0.5 g Ponceau S red
- 3 g Trichloroacetic acid
- Ad 100 ml dH$_2$O.

Pre-hybridization Solution
- 1.25 ml 20x SSPE
- 0.5 ml 100x Denhardt's solution
- 0.25 ml 10 % SDS
- 2.5 ml 100% formamide
- 0.5 ml DEPC dH$_2$O

Resolving gel (15% SDS)
- 1.2 ml dH$_2$O
- 2.5 ml Rotiphorese-Gel 30
- 1.25 ml 1.5 M Tris/HCl, pH 8.8
- 50 µl 10% SDS
- 7.5 µl TEMED
- 15 µl 10% APS

Saline-Sodium Citrate (SSC) buffer, 20x
- 17.53 g NaCl
- 8.82 g Sodium Citrate
- Ad 100 ml dH$_2$O
- Adjust to pH 7.0 with 1M HCl

Saline Sodium Phosphate-EDTA (SSPE) buffer, 20x
- 17.53 g NaCl
- 2.76 g NaH$_2$PO$_4$
- 0.74 g Na$_2$EDTA
- Ad 100 ml dH$_2$O
- Adjust to pH 7.4 with 10 M NaOH

Stacking gel
- 1.55 ml dH$_2$O
- 300 µl Rotiphorese-Gel 30
- 625 µl 1.5 M Tris/HCl, pH 6.8
- 25 µl 10% SDS
- 3.75 µl TEMED
- 7.5 µl 10% APS

Transferbuffer for western blotting
- 3.03 g Tris
- 14.41 g Glycine
- 200 ml Methanol
- 10 ml SDS
- Ad 1 l dH$_2$O

Tris-Acetat-EDTA (TAE)-buffer, 50x
- 24.2 g Tris
- 5.71 ml Glacial acetic acid
- 10 ml 0.5 M EDTA
- Ad 100 ml dH$_2$O

Tris-Borat-EDTA (TBE)-buffer, 10x
- 10.8 g Tris
- 5.5 g Boric acid
- 0.93 g EDTA
- Ad 100 ml dH$_2$O

2.1.5 Primers

Name	Sequence (5' – 3')	Melting Temp. [°C]
AdapPolyT	GAC TCG AGT CGA CAT CGA (T)17	62.4
AdapPolyC	AGC CGA TGA TCG GTA AGC (C)17	> 75
Adap1T	TCG ATG TCG ACT CGA GTC	56

Adap2C	GCT TAC CGA TCA TCG GCT	56
β-ker_uni_fw	CTG TGG TCC ATC CTG CGC TGT	63.7
β-ker_uni_rv	GCA CAT GGA GTG TTG CCT CCA AC	64.2
EcoRIadap1	AAT TGC ATT AGG CAA TGA CCT GGC G	63.0
EcoRIadap2	CGC CAG GTC ATT GCC TAA TGC	61.7
Arbitrary1	AGG GTG CCA ACC TCT TCA AGA TAC GTA	65.0
Arbitrary2	TAC GTA TCT TGA AGA GGT TGG CAC CCT	65.0
β-ker_mp_fw	AAC ATT CCA GCA ATC TTG GCA	55.9
β-ker_mp_rv	AGC CAG TGC GCC AGC ACC ATA	63.7
β-ker_pro_fw	CAT TCA GAG ATA AAT AAG GCT GTT CTT CA	61.0
sc_β-ker_sta_fw	ATG GCT GCT TGT GGT ACA TCT TGC	62.7
eu_β-ker_sta_fw	ATG GCT GCT TGT GGT CCA TCT TGC	64.4
β-ker_ter_rv	CTA ATA ACA AAT GTT ACC GCG ACG CC	63.2

All primers were ordered at Eurofins MWG Operon (Ebersberg, Germany)

2.1.6 Animals and skin samples

Moulted skin of the sandfish (*Scincus scincus*) and the Berber skink (*Eumeces schneideri*) were obtained from own specimen (purchased from official merchants). The exuviae were stored UV protected in brown glass covered with aluminium foil. Species, which were preserved in ethanol were obtained from Zoological Research Museum Alexander Koenig, Adenauerallee 160, D-53133 Bonn, Germany. Skin samples used for SEM imaging, radiation harvesting and proteomics of these preserved species were obtained by removing the outer oberhautchen using a pincer. DNA analysis was performed by isolating genomic DNA, obtained from buccal swabs with a cotton stick of living specimen (*S. scincus* and *E. schneideri*) and by dissection of liver and muscle tissue of a preserved specimen (*Meroles anchietae*). A naturally died own sandfish specimen was used for the isolation of mRNA for rapid amplification of cDNA ends (RACE).

2.2 Surface investigations

2.2.1 Comparative friction angle and abrasion resistance measurements

Measurements of the friction angles were performed with museum specimen by tightening each animal on a see-saw: the rectangular plate with the animal can be rotated on an axis between 90° (vertical) and 0° (horizontal). Sand was poured through a funnel from above on

Materials and Methods

the animal and the plate was slowly rotated into horizontal direction. The angle at which the sand began to stop gliding off from the animal's skin was determined as the friction angle (fig. 2.1). Three sandswimming species (*Scincus scincus, Scincus albifasciatus* and *Meroles anchietae*) and four non-sandswimming species (*Pantherophis guttatus, Podarcis siculus, Eumeces schneideri* and *Scincopus fasciatus*) were investigated. Data for *S. fasciatus, E. schneideri, M. anchietae, S. scincus* and *S. albisfasciatus* were investigated by Saxe, 2008 and comnbined with measurements from *P. guttatus*, and *P. siculus*, which were performed in this study. Friction angles were investigated from two specimen of each species, by measuring five friction angles from two different body locations (dorsal and ventral) of each specimen. In total, for each species 20 friction angles were evaluated.

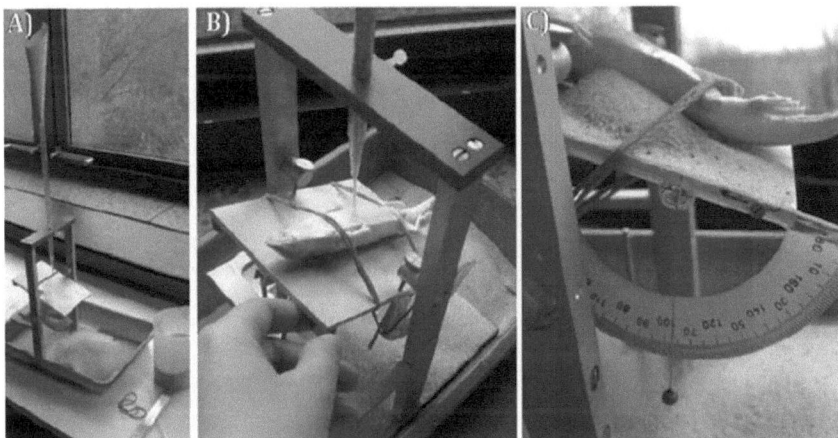

Figure 2.1 – See-saw for friction angle measurements. The see-saw (A) has a funnel with a pipette in which sand is filled. The preserved reptile was tightened on a plate (B) which can be rotated on an axis. The angle at which the sand stops to glide off from the animal can be read on an attached goniometer (C) (Saxe, 2008).

2.2.2 Scanning electron microscopy

Images of reptile skin surfaces were made either from exuviae or from scales of museum specimen. Scales were taken off with a tweezer from museum specimen and after drying on silica-gel sputter-coated with gold to a layer thickness of 300 Å. Exuviae were sputter coated without further treatment. The gold coated samples were observed using scanning electron microscopy (SEM). Images were digitally recorded using an attached i-scan digitizer with an image acquisition time of 50s. Images from *Scincopus fasciatus* (dorsal and ventral), *Eumeces schneideri* (dorsal) and *Scincus scincus* (dorsal and ventral) were investigated by Schmied,

2007. Images from *Meroles anchietae* (dorsal and ventral) and *Eumeces schneideri* (ventral) were investigated in this study.

2.2.3 Atomic force microscopy

Atomic force microscopy (AFM) images were made with a commercially available AFM equipped with a 100 nm liquid scanner. All measurements were performed in air on dry samples. For imaging of the topography, the AFM was equipped with high reflection coated triangular silicon-nitride cantilevers with a nominal spring constant of 0.05 N m^{-1}, a four sided pyramidal tip with a tip angle of 60°. Scanning was performed in contact mode including force modulation (modulation amplitude with 1.5 nm). The force setpoint was adjusted to 25 nN and images were obtained at scanning velocity of 20 μm s^{-1}. Force distance cycles were performed employing the force setpoint as used for scanning. A total retrace distance of 1 μm was chosen and the velocity was adjusted to 1 μm s^{-1}.

Adhesion force in retreat movement of the cantilever was measured for untreated glass (n = 7), native scales of *E. schneideri* (n = 5) and *S. scincus* (n = 4), silanised glycans of *E. schneideri* (n = 4) and *S. scincus* (n = 5) and adsorbed glycoproteins on glass coverslips of *E. schneideri* (n = 4) and *S. scincus* (n = 5).

2.2.4 Skin replicas

Negative replicas of the skin of museum specimen of the sandfish were made with the help of dental imprinting mass (VPS Hydro), which guarantees an accurate imprint of microstructures of the skin surface. The resulting form was poured out with epoxy resin (ratio resin:hardener 10:4) to obtain a positive replica. Replicas were made from dorsal, lateral and ventral locations of the sandfish's trunk. Furthermore, a replica without skin imprint on glass as control was made. Surface characteristics of the replicas were examinded using friction angle measurements and AFM.

2.2.5 Abrasion resistance

Abrasion resistance was investigated for skin replicas and silanised glycans on glass using the see-saw. The abrasion resistance of skin replicas was tested after friction angle measurements by evaluation of the topography using the AFM (the replicas were shortly subjected to sand during the measurements). Abrasion resistance of the covalently bound carbohydrates on glass was tested by pouring sand under two different conditions on a sample: the samples were subjected to sand that was poured on the surface either from 10 cm height for 5 minutes or from 30 cm height for 1 hour. Afterwards, the roughness of the topography was investigated on different locations (n = 6) of each sample by AFM.

2.2.6 Other possible functions of dorsal serrations

Radiation harvesting

Black construction paper was cut to exactly fit in a quartz cuvette. A hole with a diameter of 1 mm was drilled in the center of the bottom part of the cut paper (5 mm above the edge). The paper was placed on the front side of the cuvette and the zero-value calibrated photometrically by a sprectralphotometer, by measuring the wave length spectrum between 190-1100 nm. Afterwards, the outer oberhautchen of a single scale was removed from a preserved *S. scincus* specimen. The oberhautchen was carefully taped on the construction paper, covering the drilled hole. Measurements of percentage transmission in comparison to the zero calibration were carried out with a dorsal and a ventral scale in the wavelength spectrum between 190 and 1100 nm (ultraviolet to near infrared range).

Moisture harvesting

Replicas of the dorsal trunk of a sandfish as well as a smooth control (replica of glass surface) and a rough control (smooth replica of glass surface was grinded with sandpaper) were cut into pieces of exactly 15 mm in diameter. The samples were initially weighted and equilibrated to room temperature (20.5 °C). Afterwards they were held in a moisture saturated atmosphere of 80 °C for 5 seconds. The weight increase was determined immediately.

Neutralization of triboelectric charges

Pulverized polyvinylchloride (PVC) was poured through a funnel on the dorsal trunk of a preserved sandfish to find out whether triboelectric effects play a role in the sandfish. PVC is negatively charged (and thus an electron acceptor) and through the friction between the skin and the pulverized PVC, electrons should be donated by the serrations to the PVC, what in the end results in adhesion of the PVC on the skin.

2.3 DNA analysis

A fragment of the β-keratin gene from *S. scincus* was sequenced with primers derived from conserved DNA regions within known β-keratin genes. The primers were chosen by comparing β-keratin sequences of *Pantherophis guttatus* (EMBL genebank (http://www.ebi.ac.uk) accession number: AM404184), *Hemidactylus turcicus* (AM263206) and *Podarcis siculus* (AJ890445) (Valle et al., 2007). Universal keratin primers chosen were b-ker_uni_fw and b-ker_uni_rv; which was resulting in a fragment containing 309 bp (Saxe, 2008). Nucleotid blasting (http://blast.ncbi.nlm.nih.gov) of the sequence resulted in high similarity with a β-keratin coding gene of *Podarcis siculus* (AM259048).

In order to find out the complete β-keratin coding gene, different methods of DNA ligation and amplification with semi specific primers were performed.

2.3.1 General methods

DNA alignment and comparison

DNA alignment and comparison was performed with CLC Sequence Viewer 6 (CLC bio A/S).

Agarose gel elektrophoresis

1 % agarose (w/v) was added to 1x TAE buffer and dissolved by heating in a microwave until cooking. 5 µl Sybr Safe DNA Gel stain was mixed with 50 ml dissolved agarose / TAE and poured into an elektrophoresis chamber. A comb was added and the solution was allowed to cool, forming a solid gel. Thereafter, the comb was removed and the gel covered with 1 % TAE. 5 µl DNA of each sample was mixed with 1µl 6x loading buffer and applied into the wells, formed by the comb. A DNA weight marker was added into another well. The DNA was separated at about 100 mV for 45 minutes. Without further treatment, DNA was visualized with a UV-transilluminator and documented with a camera.

Freeze and squeeze

If DNA bands were needed to be sequenzed or cloned, a clean scalpel was used to cut out the desired band of the gel. Commercial available kits (e.g., agarose gel DNA extraction gel, Roche) were not suitable for short bands produced by β-keratin coding DNA. For this short DNA (< 500 bp) the freeze and squeeze technique was applied: agarose pieces containing the desired DNA were wrapped in parafilm and freezed for 30 minutes at – 20 °C. Afterwards the frozen agarose was squeezed between two fingers and the drained liquid collected in a microcentrifuge tube.

Cloning and sequencing

Desired DNA was cloned into pCR®II-TOPO® vector, transformed into DH5 alpha bacteria strain, selected by blue-white screening and sequenced (to ensure that the correct DNA fragment was transformed) by Raphael Souer as service from Fraunhofer IME, Forckenbeckstr. 6, D-52074 Aachen, Germany. Transformed bacteria that turned out to have a DNA of interest were stored in glycerol stocks: the bacteria colony was picked and bred in bacteria suspension (LB medium with 100 µg/mL ampiciline) at 37 °C overnight. 1 ml suspension was mixed with 200 µl 87% glycerine, vortexed and immediately frozen at -80 °C.

2.3.2 Rapid amplification of cDNA ends (RACE)

The experiment for RACE was carried out in cooperation with Prof. Dr. Göllner from the RWTH Aachen University, Institute of Biology III, Department of Plant Physiology, Worringer Weg 1, 52074 Aachen, Germany. Materials and devices were supplied by this institute. Isolation of mRNA from tissue was performed after Sambrooks, 2001a and the synthesis of cDNA (3'-Race and 5'-Race) after Sambrooks, 2001b. 3' RACE amplifies with a gene specific primer the start sequence of a gene and 5' RACE the terminus sequence (fig. 2.2). A combination of both RACE methods in one step is not possible due to the high variety of mRNA isolated from tissue.

Isolation of mRNA from tissue

Directly after recognition of the death of a sandfish, the animal was frozen at -80 °C to ensure intact mRNA. A piece of skin tissue of ca. 3 cm^2 was grinded under liquid nitrogen using a mortar. The grinded tissue was weighted and 1 ml trizol added per 100 mg material. After vortexing and incubation for 5 minutes at room temperature, 200 µl chloroform per ml trizol were added. Thereafter, the solution was vortexed again and after an incubation time of 3 minutes at room temperature, the solution was centrifuged for 15 minutes at 4 °C. DNA and cell proteins were resolved in the lower, organic phase and the RNA in the upper, aqueous phase. The supernatant (aqueous phase) was then transferred into a 1.5 ml Eppendorf tube and 500 µl isopropanol per ml solution was added. After vortexing and incubation for 10 minutes at room temperature, the solution was centrifuged for 10 minutes at 4 °C. The supernatant was removed and the remaining pellet washed with 500 µl ice-cold 70 % ethanol. After another centrifugation step for 5 minutes at 4 °C, the supernatant was removed and the pellet shortly dried. Afterwards, the pellet was re-suspended in 30 µl DEPC water and incubated for 10 minutes at 60 °C. After photometrical concentration measurement, the sample was frozen at -20 °C.

Materials and Methods

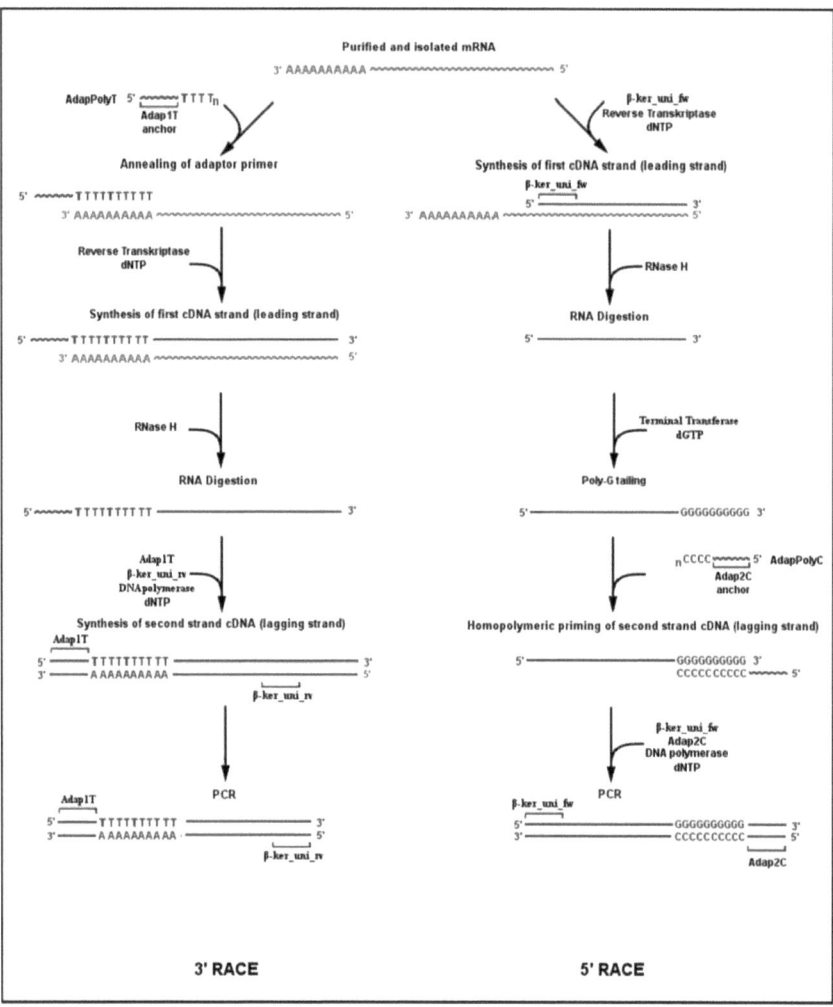

Figure 2.2 – Schematic method of RACE to amplify the start (3' RACE) and the terminus sequence (5' RACE) of a gene using amplification with anchor primers and gene specific primers.

3' RACE

An aliquot of 1 µl total mRNA was mixed with 1 µl DNase 1, 1 µl 10x Buffer and 8 µl DEPC water. The mixture was incubated for 30 minutes at 37 °C to digest all remaining DNA and afterwards incubated for 15 minutes at 70 °C to denature the DNase 1. Thereafter, the mixture was rapidly chilled on ice.

The 10 µl solution was adjusted to a total volume of 20 µl with 1 µl of 200 units per µl reverse transcriptase, 4 µl 5x reverse transcription buffer, 1 µl 20 mM dNTP solution and 4 µl (0.4 µM) AdapPolyT primer that anneals to the polyA-tail of the mRNA. The solution was first incubated at 37 °C for 60 minutes to synthesize the first cDNA strand, complimentary to the mRNA, then incubated for 10 minutes at 72 °C for 10 minutes. A positive control was carried out by qPCR using 2 µl of the mixture in a dilution of 1:15, 1 µl (0.1 µM) β-ker_uni_rv primer, 1 µl (0.1 µM) β-ker_uni_fw primer, 4 µl ddH$_2$O and 8 µl PCR Master mix Solution. PCR was run at 94 °C for 5 minutes, followed by 30 cycles of denaturation at 94 °C for 30 seconds, annealing at 58 °C for 30 seconds and elongation at 72 °C for 1 minute, with a final elongation at 72 °C for 10 minutes.

The mRNA of the cDNA/mRNA hybrid was digested by adding 0.2 µl RNase H, 10 µl 10x reaction buffer and adjusting the total volume to 100 µl with ddH$_2$O. Incubation was carried out for 2 hours at 15 °C. Afterwards, the single stranded cDNA was cleaned using High Pure PCR Product Purification Kit according to the manufacturer manual and the final solution adjusted to a volume of 10 µl.

PCR was carried out by mixing 2 µl single stranded cDNA with 2 µl (0.2 µM) Adap1T primer, 2 µl (0.2 µM) β-ker_uni_rv primer, 19 µl ddH$_2$O and 25 µl TAQ polymerase (herculase + dNTP). PCR was run at 94 °C for 5 minutes, followed by 30 cycles of denaturation at 94 °C for 30 seconds, annealing at 50 °C for 30 seconds and elongation at 72 °C for 1 minute, with a final elongation at 72 °C for 10 minutes.

5' RACE

An aliquot of 1 µl mRNA was purified from DNA as described in the protocol for 3' RACE above. The acquired mixture with a volume of 10 µl was adjusted to a total volume of 20 µl with 1 µl of 200 units per µl reverse transcriptase, 4 µl 5x reverse transcription buffer, 1 µl 20 mM dNTP solution and 4 µl (0.4 µM) β-ker_uni_fw primer, which is a gene specific primer that anneals to β-keratin mRNA. The solution was first incubated at 37 °C for 60 minutes to synthesize the first cDNA strand, complimentary to the mRNA, than incubated for 10 minutes at 72 °C for 10 minutes. A positive control was carried out by qPCR as described in the protocol for 3' RACE above.

The mRNA/DNA hybrid was purified from mRNA through RNase H digestion as described in the protocol for 3' RACE above. A poly-G tail was added to the 3' end by adding 2 µl terminal transferase, 4 µl 5x terminal transferase buffer and 4 µl 1mM dGTP solution. The final volume of 20 µl was incubated for 15 minutes at 37 °C and the transferase afterwards inactivated by heating the mixture for 3 minutes at 80 °C.

Homopolymeric priming followed by adding 4 µl (0.4 µM) AdapPolyC primer and a cleaning step was executed to remove remaining unbound primers, denatured proteins and dGTPs through the High Pure PCR Product Purification Kit and the final volume of the cDNA was adjusted to 10 µl.

PCR was carried out by mixing 2 µl cDNA with 2 µl (0.2 µM) Adap2C primer, 2 µl (0.2 µM) β-ker_uni_fw primer, 19 µl ddH$_2$O and 25 µl TAQ polymerase (herculase + dNTP). PCR was run at 94 °C for 5 minutes, followed by 30 cycles of denaturation at 94 °C for 30 seconds, annealing at 50 °C for 30 seconds and elongation at 72 °C for 1 minute, with a final elongation at 72 °C for 10 minutes.

2.3.3 RNA hybridization

This method was modified after Sagerström and Sive, 1996 as well as Saxe, 2008. A fragment of gene specific DNA was transcribed into RNA and coupled on a membrane to specifically bind genomic DNA containing the complete β-keratin gene. Before coupling, genomic DNA was restricted and ligated with arbitrary primers, annealing to the sticky end of the restriction palindrome to enable amplification. A scheme of this method is given in fig. 2.3.

Restriction of genomic DNA and ligation of palindrome specific primers

Genomic DNA, obtained from buccal swabs of living sandfishes, was purified with Geno/mini DNA Isolation Spin Kit after manufacturer's protocol. DNA concentration was measured photometrically and 500 ng used for restriction reaction. DNA was diluted to 16 µl and incubated with 2 µl 10x EcoRI buffer and 2 µl EcoRI at 37 °C overnight. On the next day, 2 µl (0.2 µM) EcoRIadap1 primer were added and the mixture heated at 65 °C for 20 minutes to denature the restriction enzyme. 6 µl 5x ligation buffer and 4 µl T4 DNA ligase were added and the solution was incubated for 10 minutes at room temperature to ligate EcoRIadap1 to the restriction palindrome. A surplus of 4 µl (0.4 µM) EcoRIadap2 primer was added, and the mixture incubated again for 10 minutes at room temperature to ligate EcoRIadap2 to the DNA. After purification with High Pure PCR Product Purification Kit the total volume of ligated DNA was adjusted to a volume of 40 µl.

Materials and Methods

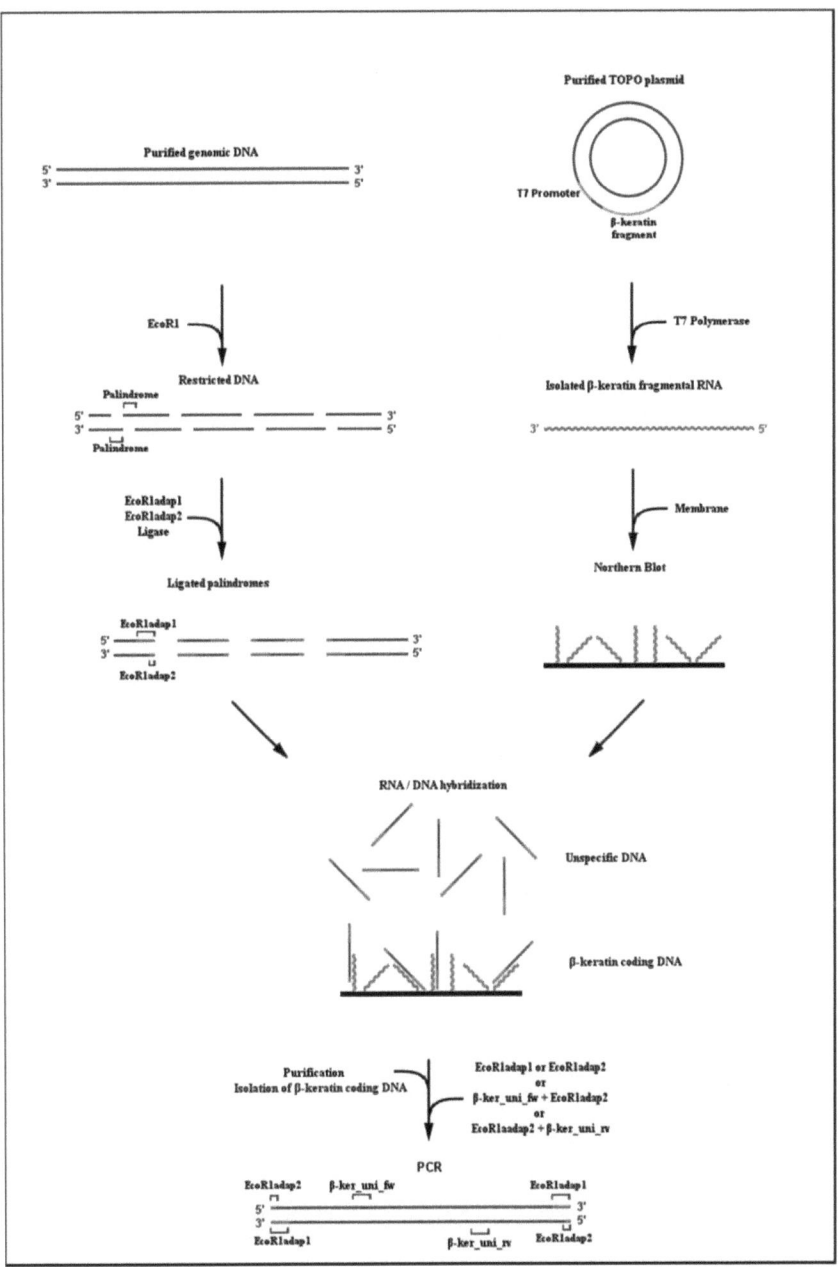

Figure 2.3 – Schematic method of RNA hybridization to purify and amplify β-keratin coding DNA.

Transcription of a DNA fragment into RNA

2 µl of genomic DNA (isolated from buccal swabs of a living sandfish) were amplified with 1 µl (0.1 µM) β-ker_uni_fw primer, 1 µl (0.1 µM) β-ker_uni_rv primer 4 µl ddH$_2$O and 8 µl PCR Master mix solution. PCR was run at 94 °C for 5 minutes, followed by 35 cycles of denaturation at 94 °C for 30 seconds, annealing at 50 °C for 30 seconds and elongation at 72 °C for 1 minute, with a final elongation at 72 °C for 10 minutes. After gel electrophoresis, DNA in the estimated range of the β-keratin fragment (ca. 300 bp) was cloned into TOPO vector.

The TOPO vector possesses a T7 expression promoter and the transcription of the β-keratin fragment into RNA was performed with the T7-transkription Kit by mixing 1 µg plasmid DNA (purified from bacteria in LB-medium with the High Pure Plasmid Isolation Kit after manufacturer's protocol), with 10 µl 10 x transcription buffer, 2 µl RNase inhibitor, 10 µl rNTP mix (10 mM each UTP, GTP, ATP and CTP) and 1.5 µl RNA polymerase. The total volume was adjusted to 55 µl with DEPC ddH$_2$O. Incubation was carried out for 2 hours at 37 °C. To ensure that transcription worked properly, 5 µl of the sample was used as control by gel electrophoresis.

Afterwards, the solution was cooled down to – 20 °C. Remaining plasmid DNA was digested by adding 2 µl DNase 1, 6 µl 10 x DNase reaction buffer and 2 µl DEPC ddH$_2$O to the 50 µl from the former step. The solution was incubated for 15 minutes at 37 °C. Thereafter, the RNA was purified from nucleotides and proteins using RNeasy mini Kit according manufacturer's protocol. Afterwards, the amount of RNA was measured photometrically.

Northern blotting of RNA

200 ng RNA was incubated in 3 volumes of a solution, containing 500 µl formamide, 162 µl 37 % formaldehyde and 100 µl 10 x MOPS buffer. The mixture was chilled on ice and mixed with 20x SSC in the ratio of 1:1. A piece of 3 x 2 cm² Hybond N membrane was cut, pre-wetted first in ddH$_2$O, then in 10x SSC. RNA was applied in aliquots of 40 µl using a dot-blot device (manufactured by the workshop of RWTH Aachen University). Afterwards, RNA was fixed by baking for 2 hours at 80 °C in an oven.

Thereafter, the membrane was incubated in 1.25 ml pre-hybridization solution for 10 minutes, and then replaced with 1.25 ml pre-hybridization solution containing 25 µl fish sperm DNA (stock solution: 1 mg/ml, sheared several times with a syringe and heated to 100

°C to denature the non-homologue DNA). The membrane was blocked in this solution for 1 hour at 42 °C.

RNA / DNA hybridization

The complete 40 µl of ligated genomic DNA was mixed with 750 µl pre-hybridization solution, 30 µl fish sperm DNA, 22 µl RNase inhibitor and 8.5 µl 1 M DTT. The membrane was incubated in this solution at 42 °C overnight. On the next day, unbound DNA was removed, by washing the membrane two times for 10 minutes at room temperature in a solution containing 2 ml 2x SSC and 0.1% SDS (w/v). Afterwards, the membrane was incubated in 1x SSC and 0.1% SDS (w/v) at 65 °C for 15 minutes.

Stripping off DNA fragments from the membrane

2 ml of 0.1x SSPE containing 0.1% SDS (w/v) were boiled and 500 µl of the hot solution poured on the membrane. The liquid was collected in a microcentrifuge tube, reboiled and poured again on the membrane. Afterwards the liquid containing the stripped off DNA fragments was cooled to room temperature and then again boiled at 100 °C for 1 minute. The DNA was cleaned using High Pure PCR Product Purification Kit and the concentration afterwards adjusted to the final volume of 20 µl. Thereafter, the DNA was ready for amplification: 2 µl of the DNA were mixed with 1 µl (0.1 µM) β-ker_uni_fw primer, 1 µl (0.1 µM) EcoR1adap2 primer, 4 µl ddH$_2$O and 8 µl PCR Master mix Solution. PCR was run at 94 °C for 5 minutes, followed by 35 cycles of denaturation at 94 °C for 30 seconds, annealing at 58 °C for 30 seconds and elongation at 72 °C for 1 minute, with a final elongation at 72 °C for 10 minutes.

2.3.4 T4 RNA ligation

This method was performed after Zhang and Chiang, 1996. A gene specific primer was used to linear amplify multiple copies of single stranded β-keratin coding DNA. T4 RNA ligase was then applied to ligate an arbitrary primer to the 3' end. PCR with the gene specific primer and the complementary arbitrary primer should amplify the gene. The schematic method is given in fig. 2.4.

Materials and Methods

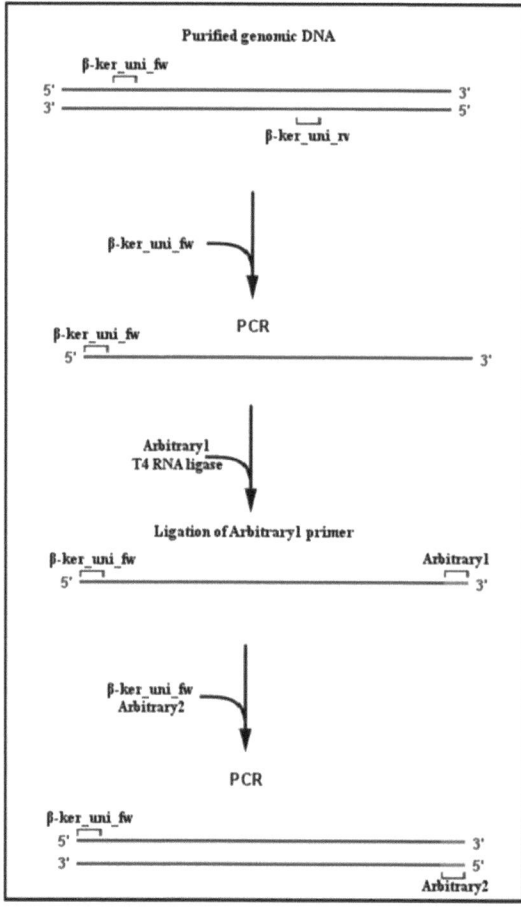

Figure 2.4 – Shematic method of T4 RNA ligation to ligate an arbitrary primer.

Synthesis of single stranded DNA

Genomic DNA, obtained from buccal swabs of living sandfishes, was purified with Geno/mini DNA Isolation Spin Kit according to manufacturer's protocol and DNA concentration measured photometrically. 200 ng DNA was mixed with 1 µl (0.1 µM) β-ker_uni_fw primer and the final volume adjusted to 20 µl with ddH$_2$O. 20 µl PCR Master mix Solution was added and PCR was run at 94 °C for 5 minutes, followed by 40 cycles of denaturation at 94 °C for 1 minute, annealing at 58 °C for 45 seconds and elongation at 72 °C for 2 minutes, with a final elongation at 72 °C for 10 minutes.

Ligation of Arbitrary 1 primer

The mixture was cleaned from DNA polymerase and dNTPs using High Pure PCR Product Purification Kit with a final volume adjusted to 16 µl. 1 µl (0.1 µM) Arbitrary 1 primer, 1 µl T4 RNA ligase (10 U) and 2µl 10x T4 RNA ligase reaction buffer was added. The mixture was incubated at room temperature overnight.

Amplification with gene specific primer and Arbitrary 2 primer

Another cleaning step was renounced due to DNA loss in this procedure. 2µl of the ligation reaction solution was mixed with 1µl (0.1 µM) β-ker_uni_fw primer, a surplus of 2 µl (0.2 µM) Arbitrary2 primer, 3 µl ddH$_2$O and 8 µl PCR Master mix Solution. PCR was run at 94 °C for 5 minutes, followed by 30 cycles of denaturation at 94 °C for 1 minute, annealing at 58 °C for 45 seconds and elongation at 72 °C for 2 minutes, with a final elongation at 72 °C for 10 minutes.

2.3.5 Amplification with semi-specific primers

In the reptile species *Anolis carolinensis*, 40 copies of β-keratin genes have been identified (Valle et al., 2008). This high number of similar β-keratin genes in a reptile species gives the possibility that two or more β-keratin genes of the sandfish are located on the same chromosome and in the vicinity of each other. In order to find out the complete gene sequence (including start and terminus sequence), the initial idea was to choose primers in the middle of the known fragment and to amplify the product in opposite direction: for example, amplification of genomic DNA with gene specific primers in opposite direction may thus result in a product containing parts of two keratin genes; one part with the start and the other with the terminus sequence. Primers in the middle of the β.keratin gene fragment (Saxe, 2008) chosen were β-ker_mp_fw and β-ker_mp_rv (fig. 2.5).

Materials and Methods

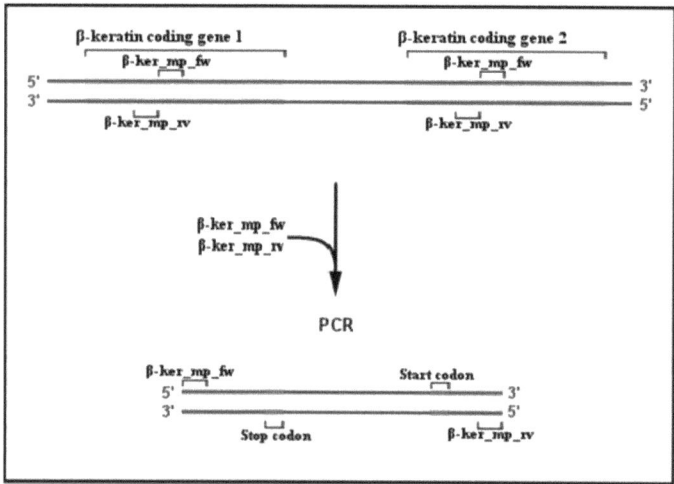

Figure 2.5 – Amplification of genomic DNA with gene specific primers in opposite direction to find the start and terminus sequence. This method presupposes that at least two similar genes are located on one chromosome.

Genomic DNA was isolated from buccal swabs from living sandfishes by Geno/mini DNA Isolation Spin-Kit according to the manufacturer's protocol. 2 µl genomic DNA was mixed with 1µl (0.1 µM) of each primer (β-ker_mp_fw and β-ker_mp_rv), 4 µl ddH2O and 8 µl PCR Master mix Solution. PCR was run at 94 °C for 5 minutes, followed by 35 cycles of denaturation at 94 °C for 30 seconds, annealing at 58 °C for 30 seconds and elongation at 72 °C for 1 minute, with a final elongation at 72 °C for 10 minutes.

The selected middle primers bound not in coding regions of two different genes (and thus amplification did not result in one sequence containing both, the start and terminus sequence). However, amplification with both primers (β-ker_mp_fw and β-ker_mp_rv) resulted in a fragment containing the start sequence, because one primer annealed in a coding region of a β-keratin gene and the other in a non-coding region in the vicinity of the gene. The β-ker_mp_fw primer served as both, forward and reverse primer, resulting in a fragment containing the terminus sequence by annealing in a coding region of the β-keratin gene and a non-coding region in the vicinity as well. Genomic DNA, containing the start and terminus sequence in the (to the sandfish closely related) Berber skink (Eumeces schneideri) was amplified in the same way. However, to identify the start sequence the forward primer was generated with the promoter sequence (instead of β-ker_mp_fw), which was part of the fragment containing the start sequence of the sandfish's β-keratin gene: β-ker_pro_fw. The parts were aligned together to display one hybrid gene for each species. The guanidine between both gene parts was inserted after comparison with the fragment obtained from the universal primers (Saxe, 2008). The sequence map with both gene parts and primer annealing positions is given in fig. 2.6.

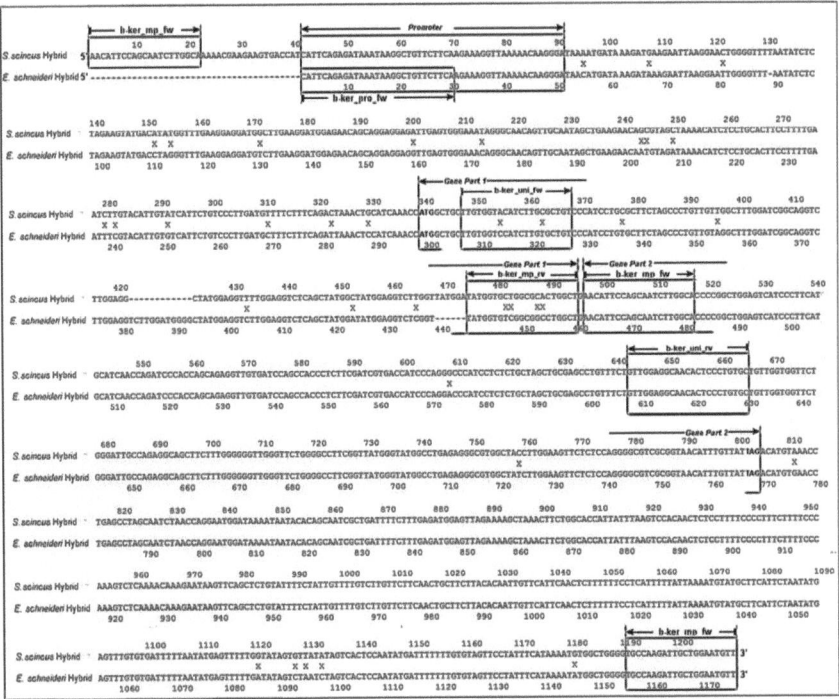

Figure 2.6 – Sequence map of two β-keratin gene parts of Scincus scincus and Eumeces schneideri. Both gene parts were aligned to form a hybrid gene with a guanidine base between (sequenced with universal primers). Next to primer annealing positions, the promoter sequence was marked and the start and terminus codon indicated with bold letters. X-signs show differences in the sequences.

With the start and terminus sequence known from both species, the sandfish and the Berber skink, complete gene amplification could be performed with sc_β-ker_sta_fw for *Scincus scincus* and eu_β-ker_sta_fw for *Eumeces schneideri*. Both species revealed the same terminus sequence; reverse primer used for amplification was β-ker_ter_rv. PCR was run with the same adjustments as for the middle primers (β-ker_mp_fw and β-ker_mp_rv) described above.

Before cloning, amplified DNA was tested to contain correct β-keratin coding DNA sequences using two control methods: (1) genomic DNA was amplified with single primers to exclude the possibility of false positive bands. An example is given in fig. 2.7, lane 1-3: Lane 1 displays amplified sandfish DNA with the primer sc_β-ker_ sta_fw alone; lane 3 with β-ker_ter_rv alone. In lane 2, amplification was performed with both primers, resulting in bands that are absent in lane 1 and 3 and in the correct β-keratin gene length of ca. 500 bp (bands are marked with a white rectangle in fig. 2.7, lane 2). The second control mechanism (2) was performed by reamplification of eluted and purified DNA (in a dilution of 1:500) with further gene specific primers. An example is given in fig. 2.7, lane 4-6: lane 4 was re-amplified using sc_b-ker_sta_fw and b-ker_ter_rv (ca. 500 bp); lane 5 was re-amplified with b-ker_uni_fw and b-ker_uni_rv, which resulted in a band of ca. 300 bp. Lane 6 was reamplified with b-ker_mp_fw and b-ker_uni_rv, which resulted in a band at ca. 150 bp. These product lengths match the sizes estimated from known β-keratin sequences. Only when both control methods produced positive results (high propability to contain β-keratin coding DNA), the amplified DNA was cloned and sequenced.

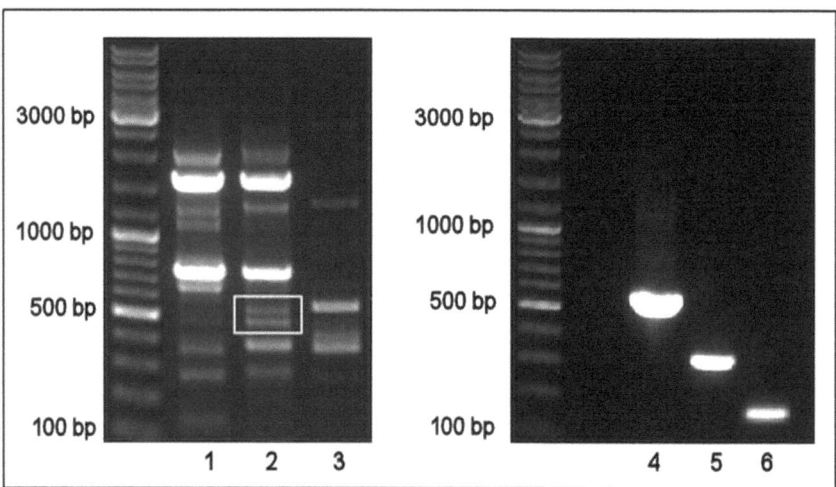

Figure 2.7 - Methods to prove that amplified genomic DNA contains a β-keratin gene. Lanes 1-3 show amplified DNA with forward primer alone (sc_b-ker_sta_fw) in lane 1 and reverse primer alone in lane 3 (b-ker_ter_rv). Lane 2 shows genomic DNA amplified with both primers. The white rectangle marks bands in the typical β-keratin range (500bp) which are absent in the two control lanes 1 and 3. Lanes 4-6 show re-amplified DNA (eluted i.e., from the white rectangle, lane 2) with different gene specific primers. Product lengths match the estimated DNA size when re-amplified with sc_b-ker_sta_fw and b-ker_ter_rv (lane 4, ca. 500 bp), b-ker_uni_fw and b-ker_uni_rv (lane 5, ca. 300 bp) and b-ker_mp_fw and b-ker_uni_rv (lane 6, ca.150 bp).

2.3.6 Sequencing of a β-keratin coding gene fragment of *Meroles anchietae*

The β-keratin coding gene of liver tissue of *Meroles anchietae* could only be sequenced using the universal primers β-ker_uni_fw and β-ker_uni_rv and thus only a fragment with ca. 150 bp missing was achieved. 2 µl isolated genomic DNA was mixed with 1µl (0.1 µM) of each primer (β-ker_uni_fw and β-ker_uni_rv), 4 µl ddH2O and 8 µl PCR Master mix Solution. PCR was run at 94 °C for 5 minutes, followed by 35 cycles of denaturation at 94 °C for 30 seconds, annealing at 58 °C for 30 seconds and elongation at 72 °C for 1 minute, with a final elongation at 72 °C for 10 minutes.

2.4 Proteomics
2.4.1 General methods

In-silico translation and alignment of β-keratin coding DNA

In-silico DNA-protein translation and alignment were performed with CLC Sequence Viewer 6 (CLC bio A/S).

Protein lysis

20 mg of the skin tissue of each species was incubated in 300µl lysis buffer at room temperature in darkness overnight. Afterwards, samples were purified from skin residues and the protein concentration was measured by a Bradford assay.

Bradford assay

Protein concentration was quantified photometrically at 578 nm using a Bradford assay (Bradford, 1976). 5 µl of protein sample was mixed with 1 ml Roti Quant and the absorption measured photometrically at 538 nm. As standard reference, 5 µl of different BSA concentrations dissolved in the lysis buffer consisting of the same chemical compounds was used.

Protein separation by SDS-gel electrophoresis

β-keratin proteins have a low molecular mass. Therefore, a 15% (v/v) sodium dodecyl sulfate (SDS) – polyacrylamide gel was chosen. The gel itself consists of two layers: the upper layer (stacking gel) with low polyacrylamide concentration serves to collect the proteins and concentrates them. The lower layer (resolving gel) separates the proteins by size. A larger and a smaller glass slide were laid together with 1 mm spacers between. The resolving gel was mixed and poured between the glass slides. After a hardening time of 30 minutes, the stacking gel was mixed and poured above the resolving gel. A comb that is forming wells, in which the proteins are applied, was inserted and the gel allowed hardening for further 30 minutes. The comb was removed and the gel transferred into a tank containing 1x electrophoresis buffer.

Samples were mixed with laemmli buffer (containing 2.5% 2-mercaptoethanol) and denatured for 5 minutes at 95°C. Each sample was applied into one well and 3 µl protein marker was given into another well.

The voltage was set to 80 V until the proteins entered the resolving gel. Then, the voltage was raised to 140 V until the blue marker band of the laemmli buffer left the gel. The stacking gel was discarded and the resolving gel used either for coomassie-blue staining or western blot.

Coomassie-blue staining of SDS-gels

The gel was transferred into a dish containing coomassie-blue staining solution. After an incubation time of 20 minutes, the gel was destained two times for 20 minutes in coomassie-blue destaining solution. Afterwards, the gel was transferred into a dish containing tab water with crepe paper and destained overnight. On the next day, only protein bands remained stained and the gel was preserved in canning foil.

Western blotting

The gel was placed on a nitrocellulose membrane and three sheets of whatman-papers added each site. The paper-gel stack was soaked in transfer buffer and transferred into a semidry blotting chamber and a currency of 44 mA was applied for 90 minutes. After blotting, the proteins on the membrane were incubated in Ponceau-red solution for 1 minute to fixate and visualize proteins. Lines were marked with a pencil. Afterwards, the proteins were destained by three washing steps in dH_2O for 10 minutes each and finally in PBS for 15 minutes.

Ink staining

For preservation after ECL visualization, the membrane was first washed in PBS Tween 0.05% for 10 minutes three times and then stained with ink staining solution for 1 hour at 37 °C. Afterwards the membrane was washed three times for 10 minutes in dH_2O and dried.

2.4.2 Visualization of β-keratin proteins through specific antibodies

20 µg proteins of a lysate of each *S. scincus* and *E. schneideri* were separated by a 15% SDS gel and blotted on a Hybond extra nitrocellulose membrane. The blots were blocked in 5% skimmed milk in PBST (w/v) overnight containing the primary antibodies in a dilution of 1:2000. Primary antibodies were obtained from Prof. Dr. L. Alibardi of the University of Bologna, Italy. β-keratin specific antibodies used were PreCB which is specific to the precorebox of β-keratins and HG-5 that is specific to a high glycan rich region of a β-keratin in *Anolis caroliensis*. A control was made by blocking the blot in skimmed milk without the addition of primary antibodies. On the next days, the blots were washed three times for 5 minutes in PBST and incubated for 1 hour with the secondary antibody goat anti rabbit peroxidase (garb-POX) in the dilution 1:3000. The blots were washed again three times for 5 minutes and the binding of the antibodies verified by the ECL system: Hydrogen peroxide, provided in the ECL-solutions, is the substrate of the peroxidase and thus oxidized so that the protons released during the reaction lead to the formation of luminal radicals, showing chemiluminescence when decaying and returning to the ground state (Rybicki and von Wechmar, 1982). 8 ml of ECL solution I was mixed with 8 ml ECL solution II and the membrane incubated in this mixture for two minutes. Afterwards, the membrane was placed in a x-ray cassette, covered with saran wrap and a x-ray film. Depending on the staining intensity the film was exposed for 1 to 4 minutes. The film was then developed, washed and fixated.

2.4.3 Visualization of glycoproteins

After separation of lysed skin proteins in a 15% SDS gel and blotting on a Hybond nitrocellulose, the nitrocellulose was blocked in 1x Rotiblock at 4 °C overnight. After blocking, the membrane was incubated for 1 hour in an incubation solution containing 3 ml PBS Tween 0.05 %, 0.5 µl concanavalin A-peroxidase and 3 µl of each 1 M $MnCl_2$, 1 M $MgCl_2$, 1 M $CaCl_2$. Afterwards, the blot was washed two times for 5 minutes in PBS to remove unspecific bounds and glycoproteins were visualized by enhanced chemiluminescence (ECL) sytem, similar to the visualization of β-keratin proteins described above.

2.4.4 Comparative glycosylation intensities

Specimen examined were sandswimming species (*Scincus scincus*, Scincidae; *Meroles anchietae*, Lacertidae) and non-sandswimming species (*Eumeces schneideri*, Scincidae; *Podarcis sicula*, Lacertidae; *Chamaeleo calyptratus*, Chamaeleonidae; *Gecko gecko*,

Gekkonidae; *Pantherophis guttatus*, Colubridae; *Pseudohajes goldi*, Elapidae; *Crocodylus suchus*, Crocodilia; *Testudo hermanni*, Testudines; *Gallus gallus* (scales), Aves; *Tyto alba* (feathers), Aves; *Homo sapiens* (hair), Mammalia).

After determination of protein quantitiy and preparation with leammli buffer, the protein samples were separated by 15% SDS-Page containing of 5 µg or an overload of 10 µg of lysed protein of each species each lane. The page was either stained with Coomassie-blue or blotted onto a nitrocellulose membrane to visualize glycosylated proteins only. Thereafter, the membrane was stained with ink. Photos were scanned with a trans-illumination scanner. Grey scale intensities in ECL-photos were evaluated using Gimp 2.6 (The Gimp Team), whereas only the film with a protein quantity of 5 µg was evaluated. The relative intensity of glycosylation was determined by marking an area within the typical β-keratin range. The mean grey-value of the area was recorded using the histogram function. The area marked was moved (same area size) to the β-keratin range of the next lane giving the next grey-value. The background was set to 0% and the grey-values from the lanes were calculated in relation to the background, what resulted in a percentile score, equivalent to the glycosylation intensity.

2.4.5 Enzymatical deglycosylation

The concentration of 8 M Urea of the proteins in the lysis buffer is too high that enzymes work properly. Therefore, the lysate was diluted 1:4 with dH_2O, because a 2 M urea concentration should not affect the function of the deglycosylation enzymes (Roche, personal information). A higher dilution factor contains the risk that too many β-keratins precipitate. After dilution, the protein concentration was measured photometrically by a Bradford assay. The positive control, asialofetuin that is both N- and O-glycosylated, was produced by dissolving the same protein quantity that resulted in the Bradford assay of the β-keratins in diluted lysis buffer.

120 µl of diluted protein samples (with each the same quantity of β-keratin and asialofetuin) were mixed with 30 µl deglycosylation buffer (containing 2 M Urea). Combinations of different deglycosylation enzymes were carried out by adjusting each sample to a final concentration of 30 µl (5 µg total protein quantity) with dH_2O, urea (to keep the final concentration at 2 M) and enzymes: 1 µl N-glycosidase F (1U) deglycosylates 20 µg protein, 1 µl O-glycosidase (1mU) as well as 1 µl sialidase (1U) deglycosylates 100-1000 µg protein.

Combinations tested for both β-keratins and asialofetuin were N-glycosidase or O-glycosidase alone, N-glycosidase together with O-glycosidase, O-glycosidase together with sialidase and finally N-glycosidase, O-glycosidase together with sialidase. Controls made were (1) diluted β-keratins and asialofetuin without deglycosylation buffer or enzymes and (2) both proteins in deglycosylation buffer but without enzymes.

2.4.6 Chemical deglycosylation through β-elimination

β-elimination is a chemical deglycosylation strategy, which is used to cleave O-linked glycans under mild conditions (Carlson and Blackwell, 1968; Lloyd et al., 1996; Morelle et al., 1998). Ammonium induced alkaline β-elimination has in contrast to sodium hydroxide the advantage that ammonium is additionally keratolytic and is thus a good alternative to the urea containing lysis buffer used otherwise, since it easily can be evaporated. Ammonium hydroxide together with moulted skin leads to keratolysis by incubation at room temperature overnight; by incubation at 60 °C overnight, the glycans get additionally β-eliminated in one step. This reaction leads to a fragmentation of proteins to peptides, which cannot be avoided. Glycan peeling is a side effect that can be reduced by addition of ammonium carbonate (Huang et al., 2001). After deglycosylation, protein fragments were removed using a membrane that removes peptides larger than 10 kDa and Dowex (H+ form) to remove smaller peptides and amino acids.

Keratinolysis with ammonium hydroxide

100 mg moulted skin of *Scincus scincus* and *Eumeces schneideri* were incubated in 2 ml 32% ammonium hydroxide at room temperature overnight. On the next day, skin residues were removed and the remaining liquid was applied on a 18x18 mm² glass coverslip and allowed to dry. Ammonium and water evaporate and a β-keratin film will be left behind that adsorbed on the glass surface.

β-elimination

The method used in this thesis was performed after Huang et al., 2001. The reaction scheme is given in fig. 2.8.

100 mg moulted skin of each *Eumeces schneideri* and *Scincus scincus* was incubated in 2 ml 32% ammonium hydroxide, saturated with ammoniumcarbonate. Additional 200 mg ammoniumcarbonate was added and incubation followed at 60 °C overnight. This reaction is highly exothermic, so that glass tubes were used. On the next day, samples were shortly stored at - 20 °C to lower the pressure inside the tubes.

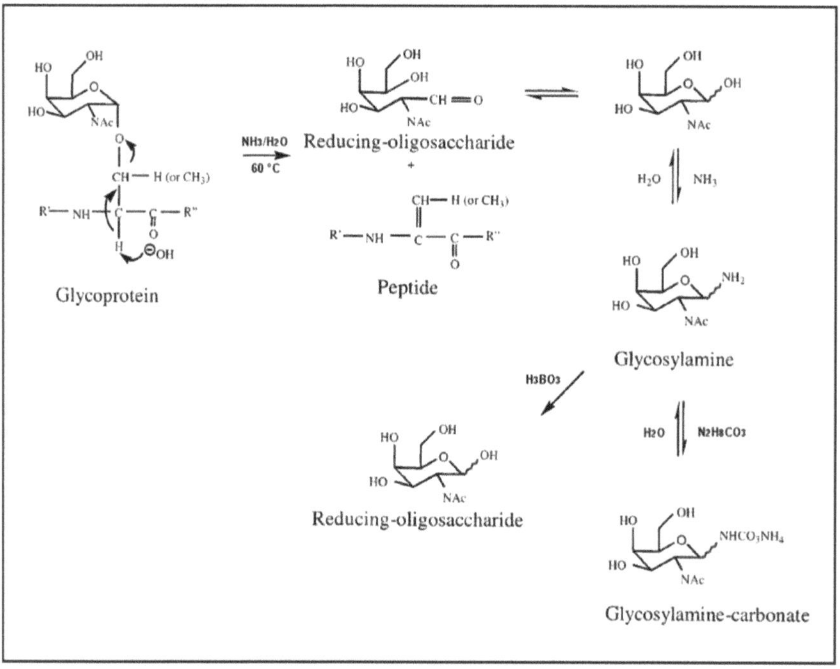

Figure 2.8 – Chemical deglycosylation through β-elimination with ammonium hydroxide (Huang et al., 2001).

Afterwards, the liquid (without skin residues) was transferred into new 1.5 ml reaction tubes and ammonium evaporated using a speedvac for 1 hour. 10 µl of 0.5 M boric acid was added to the aqueous solutions and the samples incubated at 37 °C for 30 minutes. Thereafter, the samples were completely dried using a speedvac. Afterwards, the solutions were washed by adding 1 ml methanol, followed by evaporation until complete dryness with a speedvac or under a nitrogen stream; this washing step was repeated two times (methanol forms with boric acid trimethylborate, which is highly ephemeral). Finally, the precipitate was resolved in 200 µl dH$_2$O.

Glycan purification

Glycans were purified from peptides after the β-elimination in two steps. First, larger peptides (> 10 kDa) were removed using a centiprep YM-10 tube by centrifugation of the samples for 20 minutes using this membrane. Smaller peptides were removed using Dowex 50 WX2-400 cation exchanging resin (H+ form). 150 mg Dowex was placed in an Eppendorf tube and filled up to a total volume of 1 ml with 1 M HCl. After vortexing and centrifugation, the acid was removed and replaced to a total volume of 1 ml with dH_2O. This washing step with water was repeated 3 times by vortexing, centrifugation and water replacement. Before use, the Dowex-water mixture was vortexed and then 100 µl added to each sample. After vortexing and centrifugation, the supernatant of the samples was transferred to a new Eppendorf tube, taking care to not transfer any Dowex. The samples should now contain only glycans, which can be used for silanisation or analysis. Liberated glycans however, turned out not to be very stable and peeling (hydrolysis of glycans to monosaccharids) will continue. When they are designated for silanisation, they should be processed not later than on the next day after storage at -20 °C overnight. If a MALDI-MS shall be performed they should be methylated as described in chapter 2.5.5. Peeling reactions do not affect a monosaccharide analysis.

2.5 Glycomics

2.5.1 Prediction of glycosylation sites

The prediction of N-glycosylated sites was performed with NetNGlyc (http://www.cbs.dtu.dk/services/NetNGlyc/), a prediction based on the amino acid consensus-sequence for N-linked glycans N-X-S/T. The prediction of O-glycosylation sites was performed with NetOGlyc http://www.cbs.dtu.dk/services/NetOGlyc/), based on an algorithm established through comparative studies of mucin type GalNAc O-glycosylation sites in mammalian proteins. Though this algorithm is not based on sauropsidean proteins, this algorithm takes the amino acid sequence in the vicinity of serine/threonine into account for prediction, like proline or charged amino acids (Hansen et al., 1996) and might thus be suitable for comparative prediction of O-linked carbohydrates in reptilian β-keratins as well.

2.5.2 Silanisation

Silanisation can be used to immobilize reducing glycans in a covalently bound on a glass surface, which can afterwards be used for e.g., surface investigations. Activation of the glass surface was performed with 3-Aminopropyltriethoxysilane (henceforth APTES) (Wong and Burgess, 2002). Reductive amination of the reducing terminus of the glycans were performed with the primary amine of APTES: the reaction of primary amines under acid conditions promotes with a reducing agent (e.g., sodium-cyanoborohydride) after intermediate imines the formation of secondary amines (Xia et al., 2005; de Boer et al., 2007; Song et al., 2009). A reaction scheme is given in figure 2.9.

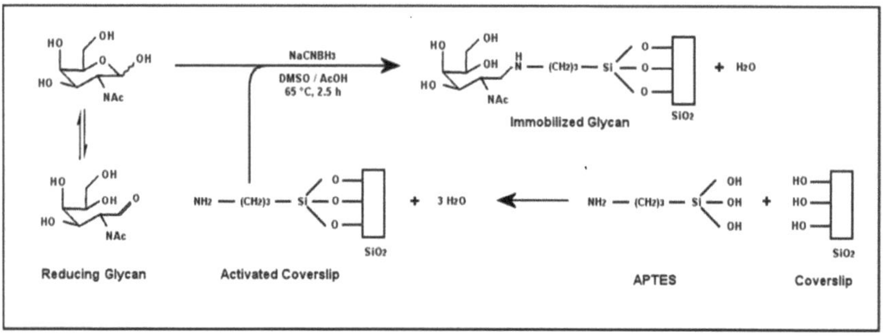

Materials and Methods

Figure 2.9 – Silanisation on glass via APTES of reducing oligosaccharides to primary amines through reductive amination (modified after Wong, 2002; Beckmann, 2010).

Reducing oligosaccharids (e.g., Sialyl-lactose and D-lactose) or purified glycans from moulted reptile skin (see chapter 2.4.6) were evaporated to a total volume of ca. 30 µl using a speedvac. The solution was mixed with 400 µl of a freshly mixed stock solution containing DMSO to glacial acetic acid in the ratio 7:3. Under a hood, 200 µl 1M sodium-cyanoborohydride was added (this chemical was added with extreme care, because mixing sodium-cyanoborohydride with acids promotes the synthesis of the highly toxic prussic acid). Three 18x18 mm² glass coverslips in petri-dishes were prepared by washing them in acetone and afterwards twice with dH_2O. The coverslips were activated with 2% APTES in acetone (v/v) for 2 minutes at room temperature. After drying at 37°C in an oven, the petri-dishes with the coverslips were transferred to a heating plate under a hood. 200 µl of the glycan-sodium cyanoborohydride mixture in DMSO/AcOH was applied with extreme care on each coverslip, so that surface tensions keep the solutions on the glass surface. The temperature of the heating plate was set to 65 °C and the coverslips incubated for 2.5 hours. Afterwards, the coverslips were washed five times for 10 minutes with PBS and two times for 10 minutes with dH_2O and then allowed to dry.

2.5.3 Surface characteristics of immobilised glycans

Surface properties of silanised glycans on glass were investigated by friction angle measurements (see chapter 2.2.1), AFM (chapter 2.2.3) and measurements of the abrasion resistance (chapter 2.2.5).

2.5.4 Quantification of linked glycans

To evaluate the quantity of immobilized glycans on a glass surface (and thus give an evidence that indeed saccharids were linked by silanisation), three coverslips silanised with glycans of the sandfish and three coverslips with glycans of the Berber skink were compared with glass controls. Coverslips with linked reptilian glycans were processed as described in chapters 2.4.6 and 2.5.2. The glass controls were treated as described in chapter 2.5.2 (activated with APTES and heated with -sodium cyanoborohydride mixture in DMSO/AcOH without addition of carbohydrates).

The coverslips were blocked in 10% Rotiblock at 4 °C overnight. On the next day, they were washed three times for 10 minutes in PBS. Afterwards, each coverslip was incubated for one hour in a solution containing 1 ml PBS Tween 0.05 %, 0.5 µl concanavalin A-peroxidase and 1 µl of each 1 M $MnCl_2$, 1 M $MgCl_2$, 1 M $CaCl_2$. After a washing step (three times for 10 minutes in PBS), 1 ml tetramethylbenzidine ready to use ELISA substrate was added to each coverslip. After an incubation time of 10 minutes, the reaction was stopped by adding 200 µl of 1M sulphuric acid. The reaction of hydrogen peroxide with peroxidise yields a blue colour with TMB as chromogenic substrate.

Stopping the reaction with acid will shift the colour to yellow, which absorption was read at 438 nm with a photometer. Quantity of linked glycans was evaluated through linked concanavalin-A-peroxidase, whereby the absorption of the glass controls was set to zero (to correct the value caused by unspecific bounds of conA). A standard curve was processed by diluting the stock solution of concanavalin-A-peroxidase (1 µg/µl) to different concentrations containing 1 ng/µl, 750 pg/µl, 500 pg/µl, 250 pg/µl and 100 pg/µl. Each 1 µl of these concentrations was incubated in 1 ml TMB for 10 minutes and measured photometrically at 438 nm, after stopping the reaction with 200 µl 1M sulphuric acid. The zero point was set by incubating 1 ml TMB without addition of conA. The quantity of linked glycans was evaluated with the absorption value using the standard curve.

2.5.5 Glycan analysis

The following methods were performed in cooperation with Prof. Dr. Franz-Georg Hanisch from the University of Cologne, Medical Faculty, Center for Biochemistry, Joseph-Stelzmann-Strasse 52, D-50931 Cologne, Germany.

In-gel pronase digestion

After skin lysis, proteins were separated with a 15% SDS-gel and stained with Coomassie-blue (see chapter 2.4.1). Desired bands within the β-keratin range at ca. 20 kDa were cut out using a scalpel. The gel stripes were washed 1 time for 10 minutes in dH_2O, afterwards 10 minutes in dH_2O:acetonitrile in the ratio 1:1 and finally for 10 minutes in acetonitrile only. After each washing step, the bands were completely dried using a speedvac. Thereafter, 100 µl ABC-buffer (50 mM ammoniumhydrogencarbonate, 30% (v/v) acetonitrile in dH_2O) were added to each stripe together with 50 µl pronase working solution (10% (w/v) pronase, 1 mM $CaCl_2$, 50 mM ammoniumhydrogencabonate in dH_2O). Digestion took place at 37 °C overnight. On the next day, digested proteins were eluted from gel with 200 µl dH_2O and afterwards, with 200 µl dH_2O:acetonitrile in the ratio 1:1. Both eluates were united and acetonitrile was evaporated using a speedvac.

Deglycosylation

Aliquots of pronase digested peptides were treated in two different ways: (1) N-linked glycans were deglycosylated with 0.5 µl PNGase F in 20 µl 50 mM ammoniumhydrogencarbonat at 37 °C overnight and separated from peptide residues by C18 solid-phase extraction; (2) to deglycosylate O-linked glycans, peptides were completely dried and resolved with 20 µl 50 mM NaOH and 0.5 M $NaBH_4$ in dH_2O under argon. Incubation took place at 50 °C overnight. Reaction was stopped on the next day with the addition of 100 µl ice-cold glacial acetic acid. Peptides were removed by addition of 50 µl Dowex 50Wx8 H^+ form. After slow rotation of the samples at room temperature for 5 minutes and followed centrifugation, the supernatant was carefully transferred (to not take up any Dowex) in a new Eppendorf tube. The sample was washed with dH_2O and evaporated to complete dryness with a speedvac. 50 µl of 1% (v/v) acetic acid in methanol was added and evaporated at 40 °C under a stream of nitrogen. This washing step was repeated 4 times.

Methylation and MALDI-MS

Samples were dried in a desiccator over KOH/P_2O_5 for 1-2 hours at room temperature. The desiccator was flooded with argon and 1 volume NaOH in dried DMSO was added. Incubation took place for 30 minutes at room temperature. Samples were frozen by addition of 0.5 vol methyliodide (cooled to -20 °C) and incubated for 30 minutes at room temperature. Thereafter, 300 µl chloroform was added, followed by neutralization with 40 µl acetic acid. Samples were washed three times by addition of 200 µl dH_2O by removing the aqueous phase

Materials and Methods

after each washing step. Finally, chloroform was evaporated under a stream of nitrogen. The pellet was resolved in 20 µl methanol and 20 µl HCCA (saturated solution of α-cyano-4-hydroxycinnamic acid in acetonitril:0.1% trifluoroacetic acid in the ratio 2:1) added. Glycan-analysis was performed with a UltrafleXtreme MALDI-TOF mass spectrometer (Bruker-Daltonik, Bremen, Germany) in positive-ion modus.

2.5.6 Monosaccharide classification

This method was performed as a service from Prof. Dr. Hanisch (University of Cologne). After deglycosylation and purification from peptide residues (see chapter 2.4.6), monosaccharids were liberated through methanolysis.

The glycans were treated in 100 µl of dry methanolic HCl (1M) containing an admixture of acetic acid methylester at equal concentration (16h at 70°C). After evaporation of methanolic HCl in a stream of nitrogen, re-N-acetylation was performed in a mixture of 100 µl methanol, 10 µl pyridine and 10 µl acetic acid anhydride (15 minutes at RT) and the monosaccharides were trimethylsilylated with 50 µl N-methyl-N-trimethylsilyltrifluoroacetamide (MSTFA) for 10 minutes at 70°C.

The monosaccharide composition was determined by GC/MS of the trimethylsilylated (TMS) 1-O-methylglycosides (Merkle and Poppe, 1994) on a Fison MD800 GC/MS (Thermo Fisher, Dreieich, Germany), equipped with a 15 m RTX5-SILMS column (Restek, Bad Homburg, Germany). After an isothermal phase at the initial temperature (100 °C, 1 min), a gradient of 6 °C/min up to 260 °C was used. MS spectra were registered by electron impact ionisation at 70eV. Relative masses between m/z 100 and 700 were scanned every second at 400 V.

2.5.7 Glycopeptide analysis

This analysis allows the determination of glycosylation sites in a known gene and the determination of glycosylation type (N- or O-linked glycans) through ammonium ion classification. The method was performed as a service from Prof. Dr. Hanisch (University of Cologne). Bands with desired β-keratins were cut out of a 15% SDS gel (Coomassie-blue stained) and protein digestion was performed with trypsine and V8-protease. Peptide analysis was carried out using a HCT ETDII PTM Discovery ESI-Ionenfallen-Massenspektrometer (Bruker-Daltonik, Bremen, Germany).

3. Results

3.1 Surface investigations

3.1.1 Comparative friction angle measurements

In order to compare the friction angle in relation to the scales' micro-structures between sandswimming and non-sandswimming species, friction angle measurements of the sandswimming species *Scincus scincus*, *Meroles anchietae* and *Scincus albifasciatus* and the non-sandswimming species *Pantherophis guttatus*, *Podarcis siculus*, *Eumeces schneideri* and *Scincopus fasciatus* were performed. The non-sandswimming species *Eumeces schneideri* and *Scincopus fasciatus* are closely related to *Scincus scincus*. Sandswimmers revealed a low friction angle between θ = 19.5° - 21.5° (corresponding friction coefficient μ = 0.35 – 0.39), whereas the friction angle of non-sandswimming species was higher θ = 26.2° - 32.2° (μ = 0.49 – 0.63) (fig. 3.1).

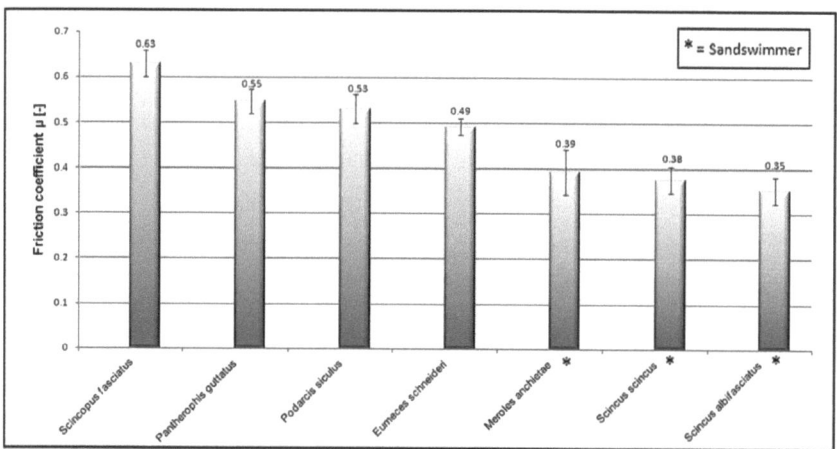

Figure 3.1 – Friction angle measurements of sandswimming (marked with asterisks) and non-sandswimming species. The friction coefficient of sandswimming species is significantly lower than the friction angle of non-sandswimming species examined as found by one way ANOVA ($P < 0.001$). Data for *S. fasciatus*, *E. schneideri*, *M. anchietae*, *S. scincus* and *S. albisfasciatus* were investigated by Saxe, 2008.

The difference between the friction angle of sandswimming and non-sandswimming species was highly significant ($P < 0.001$; one way ANOVA, $n = 140$). For pair-wise comparison of the different species a rank test, the BWS-test (Baumgartner et al., 1998; Neuhäuser et al., 2002) was employed. It was found that the friction angles of the sandswimmers does not differ significantly ($P > 0.05$) when compared pair-wise. The non-sandswimming species, not

Results

surprisingly, show more variation. The friction angles of *Podarcis siculus* and *Pantherophis guttatus* do not differ significantly ($P > 0.05$). *Eumeces schneideri* does not show a significant deviation from *Podarcis siculus* ($P > 0.05$) and only a significant but no highly significant derivation from the values obtained from *Pantherophis guttatus* ($0.05 > P > 0.01$). *Scincopus fasciatus* however deviates significantly from all other species in the sense that the friction is much higher ($P < 0.01$) when compared pair wise.

The Friction angle measurements of preserved sandfish specimen resemble values obtained from living specimen (Rechenberg et al., 2009) and no significant difference was found between dorsal and ventral scales of *S. scincus* (BWS-test for independent data: $P > 0.05$; $n = 20$).

3.1.2 Topography investigation by scanning electron microscopy

SEM-imaging of the micro-structure revealed a similar topography between the closely related *Scincus scincus*, *Eumeces schneideri* and *Scincopus fasciatus*, with serrations on the dorsal trunk of the body and a smooth ventral side, whereas no specific micro-structure was found on both, the dorsal and ventral trunk of *Meroles anchietae* (fig. 3.2).

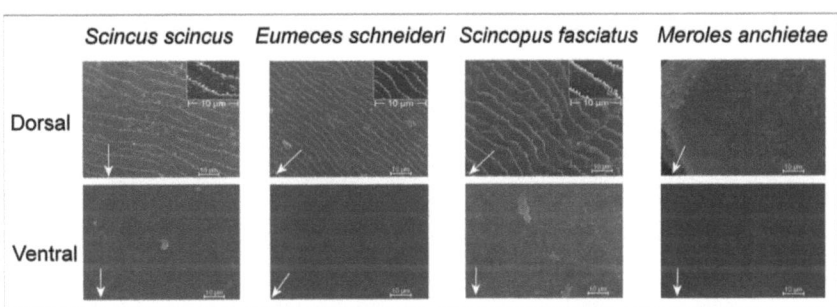

Figure 3.2 – SEM images of both ventral and dorsal sides of the trunk of two sandswimming species (*Scincus scincus* and *Meroles anchietae*) in comparison to *Eumeces schneideri*, which is closely related to *Scincus scincus*. Serrated structures are found upon dorsal scales of *S. scinus*, *E. schneideri* and *S. fasciatus* (see insert at higher magnification). All three species exhibit virtually completely smooth ventral scales. No specific pattern is found on any scales of *M. anchietae*. Cranial direction is indicated through a white arrow. Images from *Scincopus fasciatus* (dorsal and ventral), *Eumeces schneideri* (dorsal) and *Scincus scincus* (dorsal and ventral) were investigated by Schmied, 2007.

3.1.3 Surface properties of resin replicas of *Scincus scincus*

Replica imprints of ultrastructures often resembles to the original surface in properties, caused by the surface structure (e.g., Comanns et al., 2011). The replica imprints made of locations of interest of the sandfish (dorsal, ventral or lateral (fig. 3.3)) showed no difference in the

Results

friction angle from the smooth control at θ = 28° (μ = 0.53) both in caudal and cranial direction.

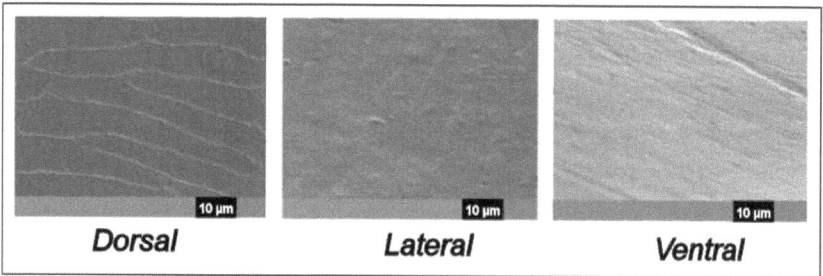

Figure 3.3 – Resin replicas of the different locations of the sandfish (SEM).

After a short time of friction angle measurements (ca. 5 minutes with a sand treatment from a height of 10 cm on the sea-saw (fig. 2.1)), strong signs of abrasion were visible on all replicas (fig. 3.4).

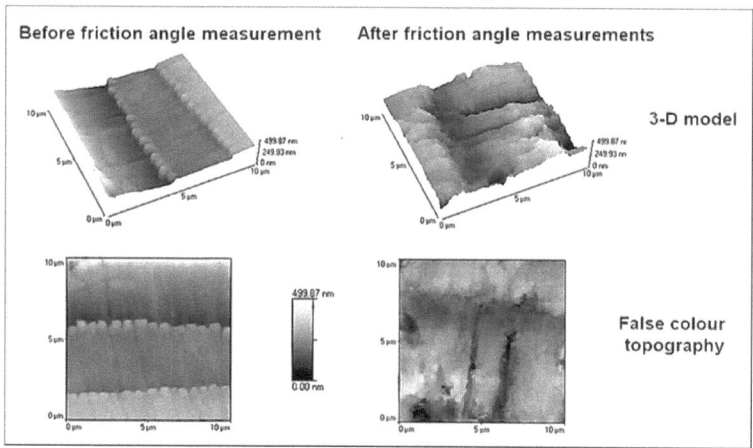

Figure 3.4 – Resin replicas before and after friction measurements on the see-saw. While the serrated structure on the replica is in good condition before the measurement, strong signs of abrasion were visible after the measurements (image was taken after ca. 5 minutes of sand treatment from a height of 10 cm on the see-saw) (AFM).

3.1.4   Other possible functions of dorsal serrations

Radiation harvesting

A possible function of the dorsal serrations is to harvest radiation of a specific wavelength, like IR-radiation. However, there was no difference in in the form of the curve of an absorbed wavelength inside the wave spectrum between 1100-190 nm (in the near infrared to ultraviolet spectrum) between dorsal scales with comb-like serrations and ventral scales that lack these serrations (fig. 3.5). The transmission is different between the scales, which is caused by different thicknesses of the scale samples. These results show that the comb-like serrations do not serve to harvest electromagnetic waves in the examined wavelengths.

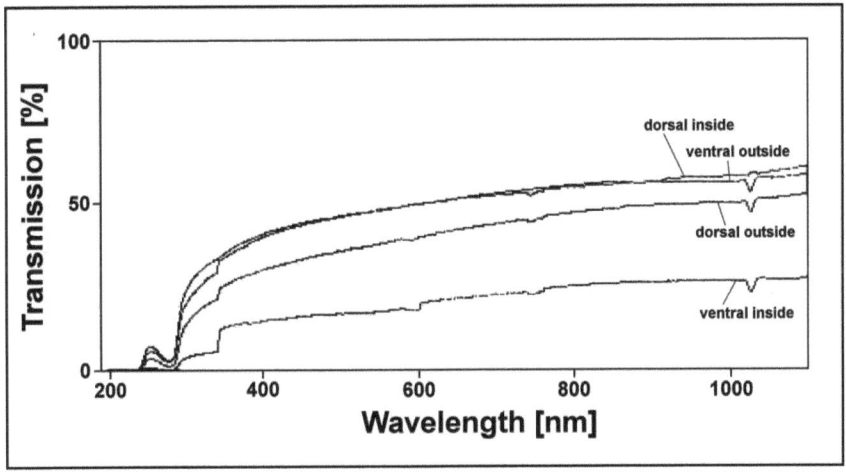

Figure 3.5 – Transmission in the IR-wavelength to UV-wavelength of moulted skin (both dorsal and ventral scales), irradiated from the inside and outside.

Moisture harvesting

In Comanns et al., 2011 it was shown that replica imprints of moisture harvesting reptiles like *Moloch horridus* have a high capacity to take up condensed water. To prove whether the dorsal serrations have similar abilities and thus serve as structures to harvest moisture, replicas of the dorsal part of the sandfish was compared with a smooth control and a rough control. The smooth replica control took up 4.00 ± 0.89 (mean $\pm$ SD; $n = 6$) mg water. The rough replica control took up 6.50 ± 1.51 ($n = 6$) mg water, and the replica of the sandfish 5.17 ± 0.75 ($n = 6$) mg water. The difference between the smooth and the rough replica was not significant (two-tailed t-test for independent data: $P = 0.07$), neither was the difference between smooth and dorsal sandfish replica (two-tailed t-test for independent data: $P = 0.09$). The difference between rough and dorsal sandfish replica was not significant as well, while the data suggests that the rough surface takes up more water than the dorsal sandfish replica (two-tailed t-test for independent data: $P = 0.09$).

Triboelectric effects

Dorsal serrations on the sandfish's trunk may serve as a lightning rod to prevent triboelectrical charges and thus an adhesion of sand on the skin surface. Therefore, pulverised PVC was poured on a reserved museum specimen. PVC (as electron acceptor) should adhere on the skin through triboelectrical charges caused by the friction when poured on the dorsal trunk of the sandfish. However, the pulverised PVC did not adhere, but glided off the skin similar to sand; means no triboelectric effect was observed.

3.2 DNA analysis
3.2.1 Rapid amplification of cDNA ends (RACE)

A qPCR of the DNA/RNA hybrid (intermediate control) did not show an amplification of any DNA, neither in 3' nor 5' RACE. Nevertheless, the protocol was carried out until the end. However, there was no cDNA amplified in final PCR as well. A possible reason for the absence of PCR products might be that the naturally died sandfish was found too late and all mRNA in the skin tissues was already completely hydrolyzed before the experiment started.

3.2.2 RNA hybridization

A β-keratin fragment (obtained by universal keratin primers) was successfully cloned in TOPO vector and transcribed into RNA by T7 polymerase (fig. 3.6).

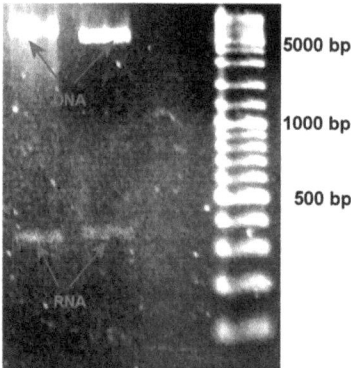

Figure 3.6 – β-keratin fragment from vector transcribed in RNA by T7 polymerase (1% agarose in TBE buffer). Plasmid DNA is marked (image was taken before DNA digestion) as well as transcribed RNA.

After northern blotting of transcribed RNA, genomic DNA with β-keratin coding DNA was selectively obtained by RNA / DNA hybridization. PCR was carried out with β-ker_uni_fw

Results

and the ligated EcoRIadap2 primer. The amplified PCR product contained one band in the estimated β-keratin length of 500 bp. This band was cloned and sequenced but did not contain the correct β-keratin sequence as found by nucleotide blast (fig 3.7).

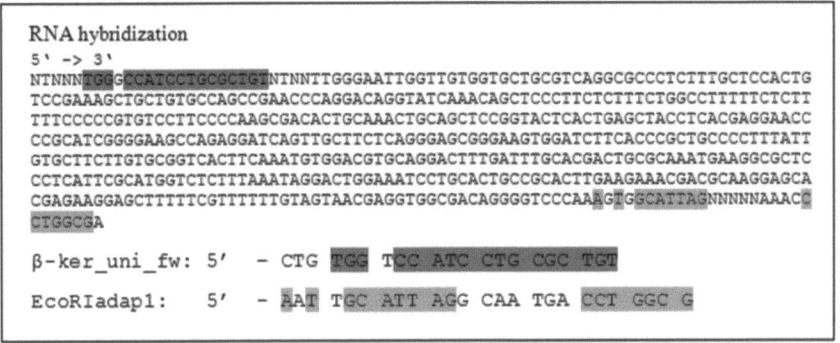

Figure 3.7 – Sequence of the PCR product after RNA hybridization, amplified with β-ker_uni_fw and EcoRIadap2 primers (EcoRIadap1 primer is complimentary to the EcoRIadap2 primer used). Both primers are (partly) present in the sequence (primers annealing sites are marked in red and turquoise). However the sequence shows no consensus with a β-keratin coding gene as found by nucleotide blast.

3.2.3 T4 RNA ligation

This method turned out to be of no use, though much effort was spent to make this easy performable protocol appropriate to find out the complete β-keratin coding gene. PCR amplified products, in which the forward primer served as both, forward and reverse primers or non-coding parts of the genome were amplified where primers annealed by chance. These problems were already reported by Saxe, 2008.

3.2.4 Amplification with semi-specific primers

The β-keratin coding genes of *Scinus scincus* and *Eumeces schnederi* were sequenced in three steps. First of all a fragment was sequenced using universal β-keratin primers derived from conserved regions (Saxe, 2008). In a second step, primers inside of this fragment were chosen that were specific to coding and non-coding regions, resulting in two gene parts. One part contained the terminus codon and the other part contained the start codon with the promoter sequence 5':
CATTCAGAGATAAATAAGGCTGTTCTTCAAGAAAGGTTAAAAACAAGGGA (score: 0.76; http://www.fruitfly.org/seq_tools/promoter.html). Both parts were aligned together forming a hybrid gene, with a guanidine base (from step 1) in between to close the gap

(sequence map is given in fig. 2.6). As final step, with the start and terminus sequence known, primers were generated to amplify and sequence a complete β-keratin gene of both species, *Scincus scincus* and *Eumeces schneideri* (sequence map is given in figure 3.8).

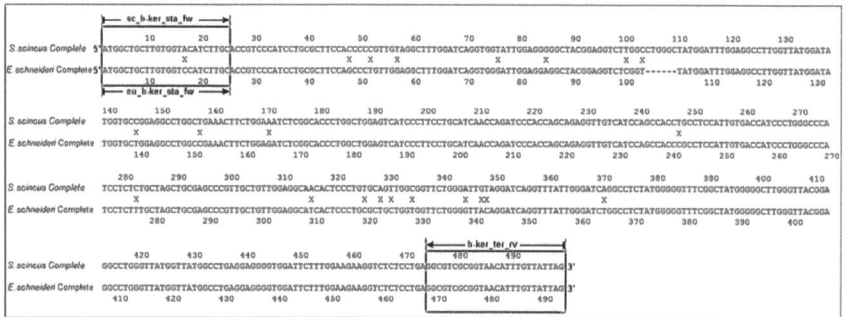

Figure 3.8 – Sequence map of complete β-keratin genes of *Scincus scincus* and *Eumeces schneideri*. Primer annealing positions and differences in the sequences (X-signs) were marked.

The DNA of both species show only little difference, indicating homologue genes sequenced in the closely related species. A nucleotide blast, performed with the complete sequences, revealed high similarity with *Podarcis siculus* gprp-3 gene (EMBL access number: AM259049): *Scincus scincus* β-keratin complete gene has a query coverage of 70% and a maximal identity of 74 %; *Eumeces schneideri* β-keratin complete gene has a query coverage of 68 % and a maximal identity of 71 %.

3.2.5 Sequencing of a β-keratin coding gene fragment of *Meroles anchietae*

The DNA of *Meroles anchietae*, obtained from liver tissue from preserved specimen turned out to be not very stable. Though experiencing difficulties, it was possible to amplify at least a fragment (where approximately 3 amino acids are missing at the start and 50 amino acids are missing at the end of the gene) using universal β-keratin primers (fig 3.9). A nucleotide blast shows high similarity with gprp-2 gene of *Podarcis siculus* (EMBL access number: AM259048.1; query coverage 100% with a maximal identity of 81%).

Results

```
Meroles anchietae (β-keratin fragment)

5' -> 3'
CTGTGGTCCATCCTGCGCTGTCCCATCCTGTGCTTCAACCCCCACCGTTGGATTTGGATCAGCAGGTGGTCTTGG
CTATGGAGTTCTTGGTAGAGGGGGAGGTCTCGGCTATGGATACGGAGGTCTCGGCTATGGTTTTGGAGGTGCTGA
AAGAGCAACCAACCTTGGAATCCTGGCAGGAGTTGTCCCATCATGTGTCAACCAGATCCCACCAGCAGAGGTTGT
GATCCAGCCACCCCCCACCGTCCTGACCATCCCAGGGCCCATCCTCTCTGCCAGCTGTGAGCCAGTGGCTGTTGG
AGGCAACACTCCATGTGC

β-ker_uni_fw: 5' - CTG TGG TCC ATC CTG CGC TGT
β-ker_uni_rv: 5' - GCA CAT GGA GTG TTG CCT CCA AC
```

Figure 3.9 – Sequence of the β-keratin fragment of *Meroles anchietae*, amplified with β-ker_uni_fw and β-ker_uni_rv primer (the forward primer is marked with red and the complimentary sequence of the reverse primer is marked in turquoise). The sequence shows high similarity with gprp-2 gene (*Podarcis siculus*) as found by nucleotide blast.

3.3 Proteomics

3.3.1 In-silico translation of β-keratin coding DNA sequences

Due to the high similarity, the translated amino acid sequences of the complete β-keratin genes of both species *Scincus scincus* and *Eumeces schneideri* (Scincidae) were also compared with *Podarcis siculus* (Lacertidae) gprp-3 gene (fig. 3.10). *Scincus scincus* and *Podarcis siculus* differ in 63 amino acids and *Eumeces schneideri* and *Podarcis siculus* differ in 70 amino acids.

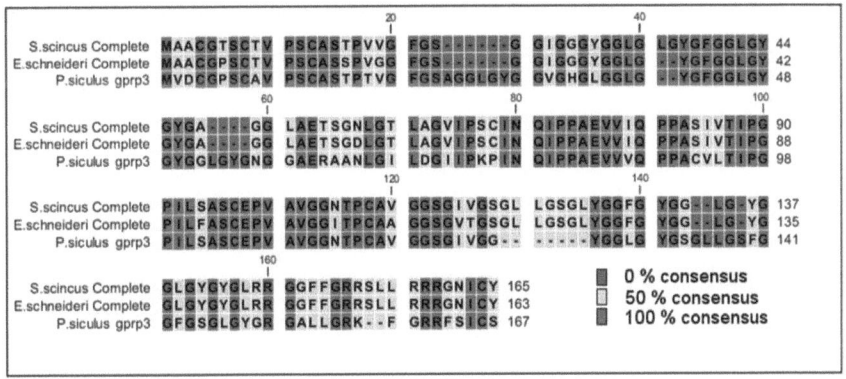

Figure 3.10 – β-keratin protein sequence alignment of *Scincus scincus*, *Eumeces schneideri* and *Podarcis siculus* gprp-3. The consensus is marked in blue for 0%, pink for 50% and red for 100%.

Between the closely related *Scincus scincus* and *Eumeces schneideri* (Scincidae) eleven of 165 amino acids are different, from which two are missing in *E. schneideri*. Of the remaining nine amino acid differences in the sandfish is one proline changed to threonine at postion 6, one serine changed to threonine at position 16 and one phenylalanine changed to serine at position 94 (fig. 3.11). Threonine and serine enable O-glycosylation (see chapter 3.4). Furthermore, 2 amino acids have been changed to asparagine in *S. scincus* at positions 57 and 105. Asparagine enables N-glycosylation however, only in the consensus Asn-X-Ser/Thr which is not part of the sequence.

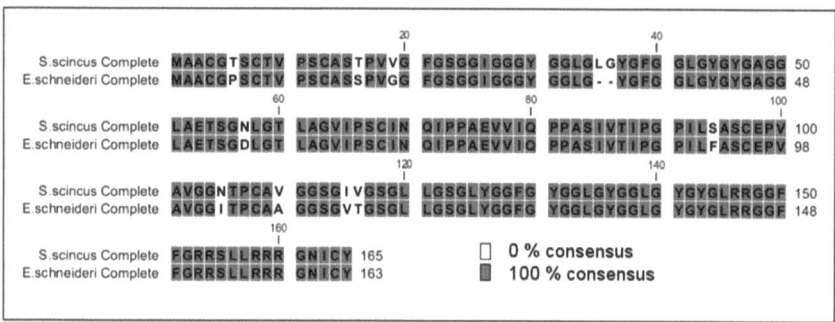

Figure 3.11 – β-keratin protein sequence alignment of *Scincus scincus* and *Eumeces schneideri*. The consensus is marked in white for 0% and red for 100%.

The β-keratin fragment of *Meroles anchietae* shows high similarity with *Podarcis siculus* gprp-2 gene (both belong to the family Lacertidae). Between both sequences, 22 amino acids are different, from which 8 are missing in *Meroles anchietae*. From the remaining 14 amino acids different in *M. anchietae*, one alanine is changed to threonine at position 64, one lysine is changed to serine at position 75 and one cysteine changed to threonine at position 92 (fig. 3.12).

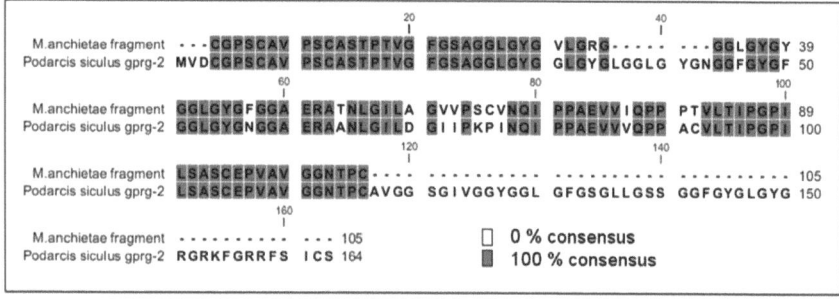

Figure 3.12 – β-keratin protein sequence alignment of *Meroles anchietae* (fragment) and *Podarcis siculus* gprp-2. The consensus is marked in white for 0% and red for 100%.

3.3.2 Visualization of β-keratin proteins through specific antibodies

Only one tested β-keratin specific primary antibody, PreCB (specific for the pre-corebox), bound on the β-keratin proteins. An antibody, which is specific to a high glycine region of β-keratins of *Anolis carolinensis* (HG-5) did not react with β-keratins of *S. scincus* and *E. schneideri*. Though 20 μg protein and a low dilution of the primary antibody led to a signal overload, the prominent β-keratin bands at 17 kDa were stained as well as bands at about 30 kDa in both species (fig. 3.13). This indicates that there are at least two types of β-keratins present in the investigated skink species.

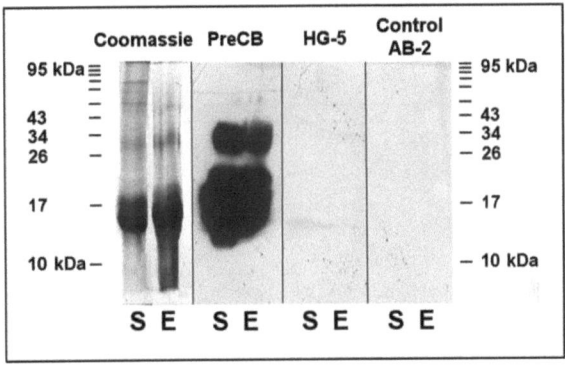

Figure 3.13 – Reactions with β-keratin specific antibodies. The Coomassie-blue stained gel visualizes all proteins present in the lysates. PreCB primary antibody is specific to the pre-corebox of β-keratin proteins and HG-5 specific to a glycine rich region of β-keratins of *Anolis carolinensis*. A control was made by incubation with the secondary antibody only. S: *Scincus scincus*; E: *Eumeces schneideri*.

3.3.3 Comparative glycosylation intensities

Glycosylated proteins of skin lysates of various sauropsidean species were visualized by western blot employing peroxydase labelled concanavalin A as detection system to find out whether β-keratins in these species are glycosylated as well and to evaluate glycosylation intensities. When 5 μg protein were applied on each lane of selected species only the glycosylated β-keratins of the sandswimming species *Scincus scincus* and *Meroles anchietae* are visible (next to *Crocodylus suchus*) (fig. 3.14). This indicates a significantly stronger glycosylation of the β-keratins in sandswimming species.

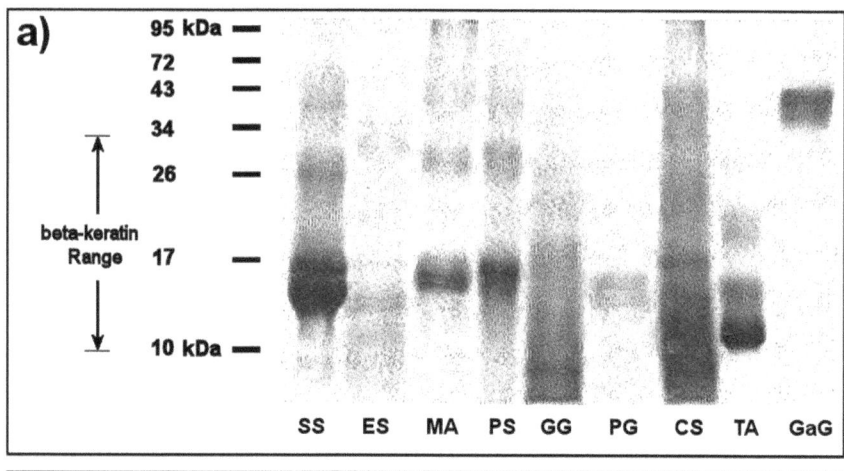

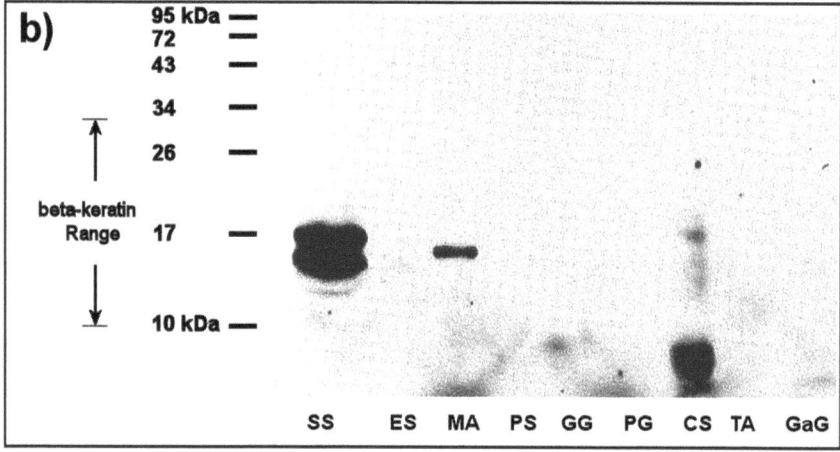

Figure 3.14 – a) Coomassie-blue stain of dissolved scales on a 15% SDS-PAGE (stains all proteins); b) western blot with glycoproteins visualized by ECL-system (stains glycoproteins only) with concanavalin A as carbohydrate marker. This figure shows a single PAGE/blot with selected species examined with 5 µg protein each lane. In b), when applying 5 µg protein each lane, only the sandswimming species *Scincus scincus* (SS) and *Meroles anchietae* (MA) show glycosylated β-keratins (next to *Crocodylus suchus* (CS)). SS: *Scincus scincus*, ES: *Eumeces schneideri*, MA: *Meroles anchietae*, PS: *Podarcis siculus*, GG: *Gekko gecko*, PG: *Pantherophis guttatus*, CS: *Crocodylus suchus*, TA: *Tyto alba*, GaG: *Gallus gallus*.

An evaluation of glycosylation intensities by mean grey-values of glycosylated bands by areas of the same size of the photo shown in figure 3.14 b (with 5 µg protein each lane), revealed a higher glycosylation intensity in the sandswimming species (fig. 3.15). The high value of *Crocodylus suchus* in comparison to *Meroles anchietae* is resulting from the large area of the weak band in the β-keratin range. The glycosylated protein below 10 kDa in *Crocodylus suchus* is most likely no β-keratin protein.

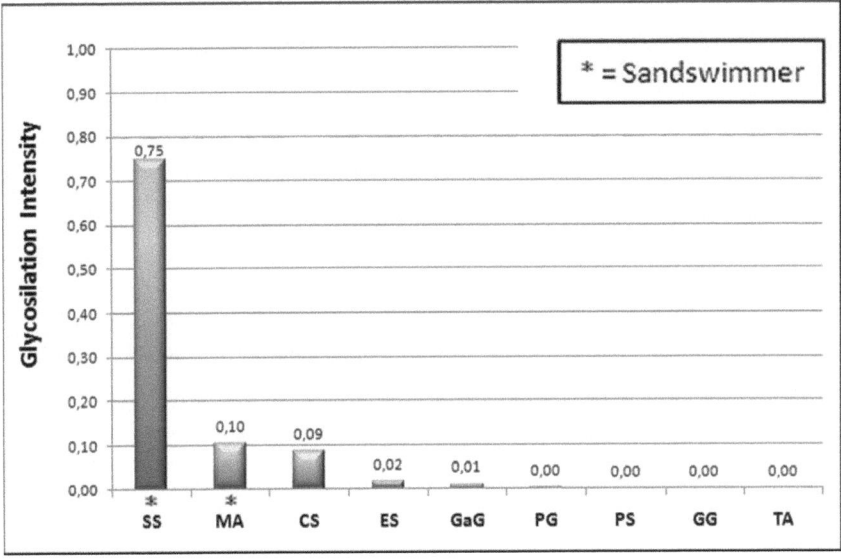

Figure 3.15 – Relative lectin staining intensities of β-keratins of selected sauropsidean species shown in figure 3b, evaluated by mean grey-values for areas within the typical beta-keratin range by a graphical program and set in relation to the background. Sandswimming species (marked with asterisks) and *Crocodylus suchus* show a band that is absent in other species at a quantity of 5 µg protein each lane. SS: *Scincus scincus*, MA: *Meroles anchietae*, CS: *Crocodylus suchus*, ES: *Eumeces schneideri*, GaG: *Gallus gallus* (scale), PG: *Pantherophis guttatus*, PS: *Podarcis siculus*, GG: *Gekko gecko*, TA: *Tyto alba* (feathers).

Proteins inside the typical β-keratin range (ca. 10 - 30 kDa) of all sauropsidean species (and also *Homo sapiens* (hair)) examined have been found glycosylated when applying an overload of 10 µg protein each lane (fig. 3.16). Films of the blots with the protein overload were unusable for exact graphical estimation.

Results

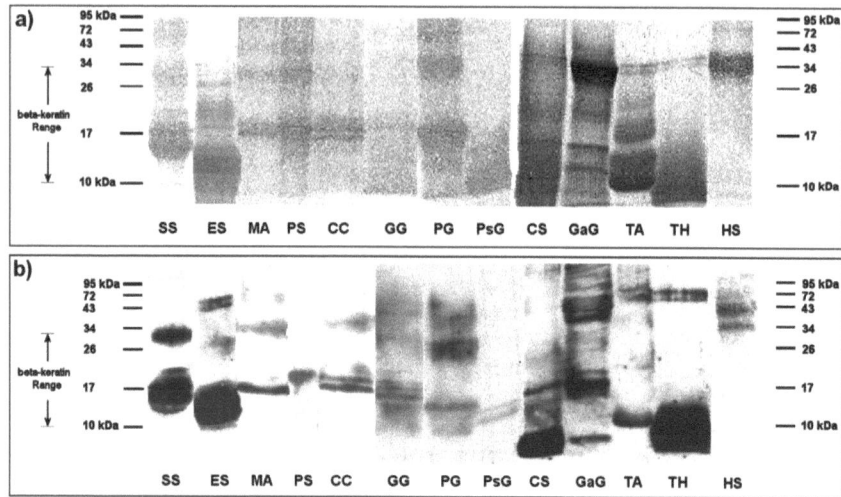

Figure 3.16 – a) Coomassie-blue stain of dissolved scales on a 15% SDS-PAGE (stains all proteins); b) western blot with glycoproteins visualized by ECL-system (stains glycoproteins only) with concanavalin A as carbohydrate marker. This figure shows merged PAGEs/blots of various sauropsidean species examined with an overload of 10 .g protein each lane. At this protein quantity all sauropsidean species show glycosylated keratins in b). *Eumeces schneideri* (ES) and *Scincus scincus* (SS) show a similar glycosylation pattern, however *S. scincus* has a higher molecule mass than *E. schneideri*. SS: *Scincus scincus*, ES: *Eumeces schneideri*, MA: *Meroles anchietae*, PS: *Podarcis siculus*, CC: *Chamaeleo calyptratus*, GG: *Gekko gecko*, PG: *Pantherophis guttatus*, PsG: *Pseudohaje goldii*, CS: *Crocodylus suchus*, GaG: *Gallus gallus* (scales), TA: *Tyto alba* (feathers), TH: *Tesdudines hermanni*, HS: *Homo sapiens* (hair).

3.3.4 Enzymatical deglycosylation

Enzymatical deglycosylation was performed to find out whether β-keratins are N- or O-glycosylated. To avoid precipitation of the keratins, lysate (as well as the positive control asialofetuin, which is both N- and O-glycosylated) was diluted to a urea concentration of 2 M and treated with different glycosidases and combinations of them (fig. 3.17).

Asialofetuin was completely deglycosylated when incubated with a combination of N- and O-glycosidase (NO) as well as both enzymes and the addition of sialidase (NSO). Incubation with one enzyme alone (N or O) led only to a weakening of the glycosylation intensity. The β-keratin intensities of the lysate (at ca. 17 kDa) were not affected by any treatment with enzymes. However, the intensity of another glycoprotein (at ca. 30 kDa), which presumably is a β-keratin as well (see chapter 3.3.2) became significantly weaker whenever N-glycosidase was applied (NSO, NO and N).

Results

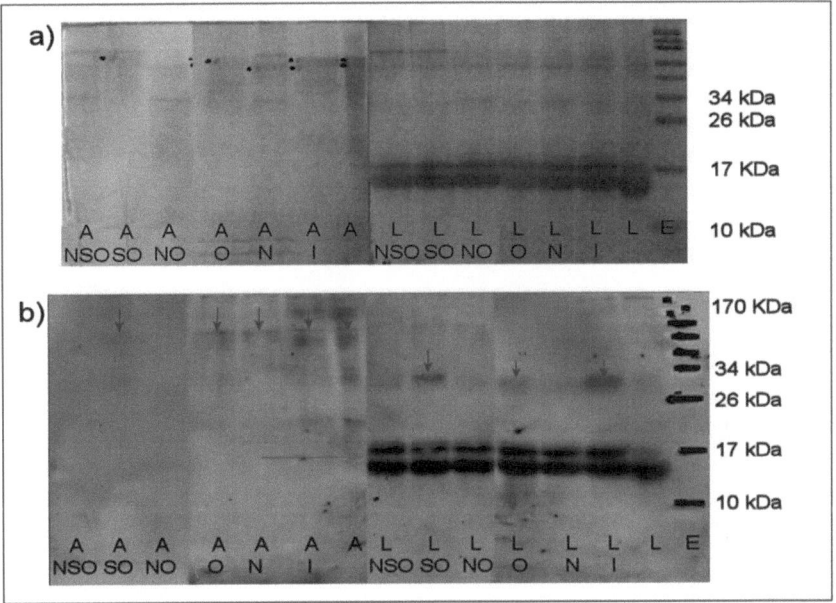

Figure 3.17 – a) Coomassie-blue stain of dissolved sandfish scales on a 15% SDS-PAGE (stains all proteins); b) western blot with glycoproteins visualized by ECL-system (stains glycoproteins only) with concanavalin A as carbohydrate marker. This figure shows merged PAGEs/blotss of asialofetuin (A; positive control) and sandfish skin lysate (L) treated with different glycosidases and combinations of them (N: N-glycosidase; O: O-glycosidase, S: sialidase, I: incubation buffer without glycosidases). The red arrow shows the glycosylated prominent β-keratins and the blue arrows other glycosylated β-keratins present in the lysate, respectively glycosylated asialofetuin. If this blue arrow is missing in a line, a deglycosylation occurred.

3.3.5 Chemical deglycosylation through β-elimination

Alkaline β-elimination was used to release O-linked glycans from skin exuviae, since enzymatical deglycosylation did not affect the glycosylation intensity of the β-keratin bands at 17 kDa as shown in fig. 3.17. Ammonium based alkaline conditions leads also to keratinolysis (which is an alternative to the urea containing lysis buffer), whereby fragmentation of protein is small when incubation occurs at room temperature. When incubation with ammonium is carried out at 60 °C, a β-elimination of the glycans is achieved; however the proteins get highly fragmentized. Peptides can be purified from liberated glycans by using Dowex (H^+-form) (fig. 3.18).

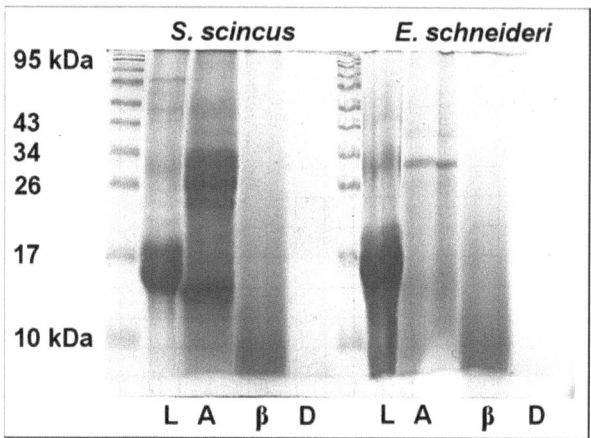

Figure 3.18 – Coomassie-blue stain of a 15% SDS-PAGE containing proteins of *Scincus scincus* and *Eumeces schneideri*. L: skin proteins lysed in lysis buffer. A: skin proteins lysed in 32% ammonium at room temperature. β: skin proteins lysed in 32% ammonium at 60 °C leads (next to alkaline β-elimination of glycans) to protein fragmentation. D: fragmentized peptides after β-elimination were removed by Dowex.

3.4 Glycomics

3.4.1 Prediction of glycosylation sites

There is no N-glycosylation (indicated through the consensus sequence Asn-X-Ser/Thr) site present in any amino acid sequence compared. The predicted O-glycosylation sites are shown in fig. 3.19 for each β-keratin sequence compared by families. The program NetOglyc predicts glycosylation sites by calculating a score, whereby the threshold for possible glycosylation is at 0.5. The exchange of different amino acids to serine or threonine in *Scincus scincus* led to an increased O-glycosylation potential at position 6 (score: 0.62), 16 (score: 0.51), 94 (score: 0.49) and 96 (score: 0.50) in comparison to *Eumeces schneideri* (marked with an arrow in figure 3.19). *Eumeces schneideri* has also one amino acid exchanged to threonine at position 114 that is not present in *Scincus scincus*, however this amino acid reaches a score of only 0.38.

The β-keratin fragment of *Meroles anchietae* has 3 serines / threonines more than the β-keratin sequence of the same length of *Podarcis siculus* (postions 64, 75 and 92), filling a gap in the sequence. However, only 1 exchange from cysteine to threonine at postion 92 has a potential reaching the threshold with a score of 0.64). Proline residues are commonly found in the sequence close to the O-glycosylated serine or threonine (Young et al., 1979), which is the case for positions 75 and 92.

Results

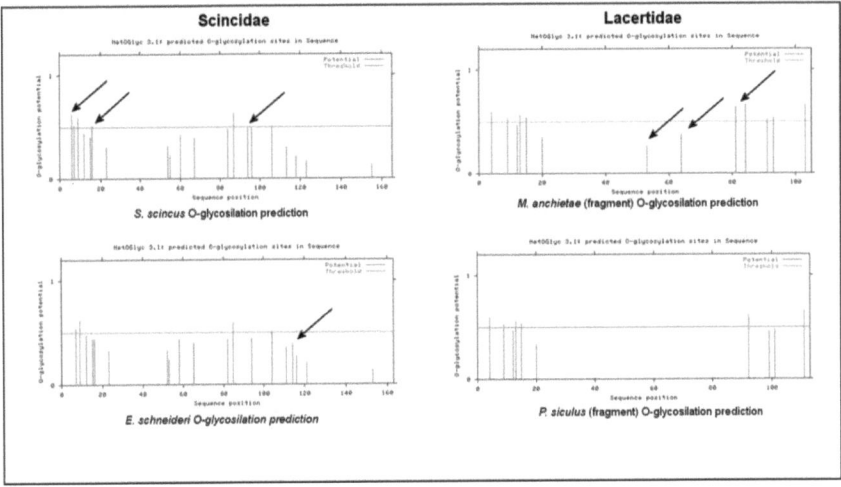

Figure 3.19 – O-glycosylation potential of exchanged serines and threonines predicted by NetOglyc in comparison to closely related sandswimming (*S. scincus* and *M. anchietae*) and non-sandswimming species (*E. schneideri* and *Podarcis siculus*). Exchanged serines and threonines are marked with an arrow. The program NetOglyc calculates a score giving a probability for glycosylation. When this score reaches the threshold of 0.5 (red line), the amino acid is possibly glycosylated.

3.4.2 Surface characteristics of adsorbed proteins and silanised glycans on glass

Adhesion force

The adhesion force of native scales, adsorbed glycoproteins on a glass surface and covalently linked glycans on a glass surface (via silanisation after β-elimination) was investigated by AFM (fig. 3.20). The samples of *S. scincus* and *E. schneideri* were compared with untreated glass. All three samples of the sandfish exhibit a similar, very low adhesion force (indicated with ΔF in fig. 3.20) by retreat movement of the cantilever in comparison to *E. schneideri*, while showing strong differences in the topography (Glass control: ΔF = 27.0 nN ± SD = 5.0 nN ($n = 7$); *Eumeces schneideri* native scales: ΔF = 16.5 nN ± SD = 3.2 nN ($n = 5$), silanised glycans: ΔF = 17.1 nN ± SD = 4.1 nN ($n = 4$), adsorbed proteins: ΔF = 15.2 nN ± SD = 3.4 nN ($n = 4$); *Scincus scincus* native scales: ΔF = 2.1 nN ± SD = 1.4 nN ($n = 4$), silanised glycans: ΔF = 2.4 nN ± SD = 1.6 nN ($n = 5$), adsorbed proteins: ΔF = 2.5 nN ± SD = 1.4 nN ($n = 5$)).

Results

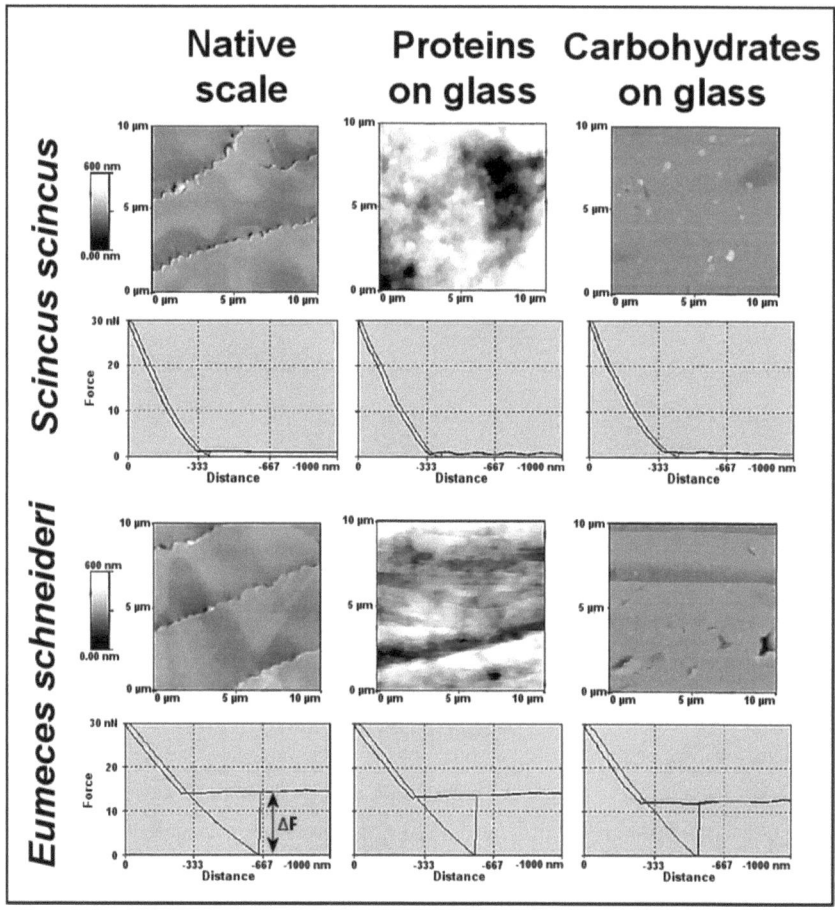

Figure 3.20 – Comparative adhesion force (investigated by AFM) of native scales, adsorbed proteins and silanised glycans of *S. scincus* and *E. schneidei*. The force difference ΔF in retreat movement (adhesive force) of the cantilever is indicated in the force-distance diagram of native scales of *E. schneideri*.

Friction angle

In order to determine whether the glycans found in the scales of *S. scincus* are sufficient to reduce sand friction, native glycans obtained from *S. scincus* scales and *E. schneideri* scales respectively, were bound to primary amines of the silane groups of activated glass coverslips. The friction angles of the silanised glycans resemble characteristics of the native scales shown in fig.3.1 The friction coefficient was compared with the adhesion force of force-distance diagrams, investigated by AFM (fig. 3.20), which indicates a correlation between friction coefficient and adhesion force (fig. 3.21). Untreated glass showed the highest adhesion force and friction angle, followed by samples of *Eumeces schneideri*. *Scincus scincus* exhibits almost no adhesion force and shows also the lowest friction angle of the examined samples (*S. scincus* native scales: $\mu = 0.38 \pm SD = 0.03$ ($n = 20$); silanised glycans: $\mu = 0.36 \pm SD = 0.02$ ($n = 20$)). Silanised glycans of *E. schneideri* resemble the natural model as well, but show a higher friction coefficient than the sandfish (*E. schneideri* native scales: $\mu = 0.49 \pm SD = 0.02$ ($n = 20$); silanised glycans: $\mu = 0.48 \pm SD = 0.02$ ($n = 20$)). A glass surface (control) has a higher friction coefficient ($\mu = 0.52 \pm SD = 0.02$ ($n = 20$)), which is in close agreement with measurements by Rechenberg et al., 2004.

The friction angle of adsorbed proteins could not be measured due to high abrasion through sand of the protein layer during measurements with the see-saw; the non-covalently bound proteins abraded through the sand after a short while. However, at the beginning of the measurements, also protein layers showed low friction behaviour similar to native scales of the sandfish.

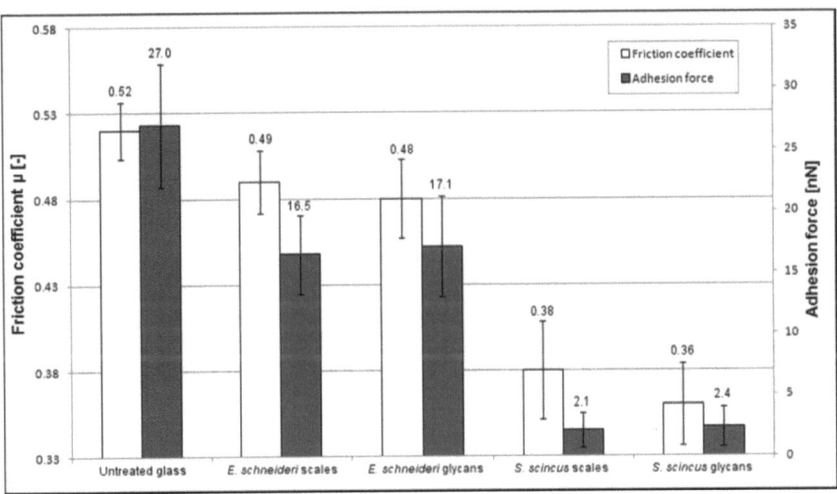

Figure 3.21 – Friction coefficient versus adhesion force. Compared was untreated glass with native scales and silanised glycans of *S. scincus* and *E. schneideri*.

Abrasion resistance

The covalently bound glycans of the sandfish on glass show a higher abrasion resistance than glycans obtained from *E. schneideri* and an untreated glass control by measuring the roughness of the surface after treatment with sandblasting for 5 minutes from a height of 10 cm (fig. 3.22); glass control: initial Ra = 1.2 nm ± SD = 0.9 nm (n = 6), after 5 minutes sandblasting Ra = 10.3 nm ± SD = 2.0 nm (n = 6); *E. schneideri*: silanised glycans initial Ra = 3.5 nm ± SD = 1.0 nm (n = 6), after 5 minutes sandblasting Ra = 9.8 nm ± SD = 1.9 nm (n = 6); *S. scincucs*: silanised glycans initial Ra = 3.7 nm ± SD = 1.1 nm (n = 6), after 5 minutes sandblasting Ra = 6.1 nm ± SD = 2.0 nm (n = 6). After treatment under harsher conditions (1 hour from a height of 30 cm), no difference in the roughness was detectable; all samples became dull and scratched (data not shown).

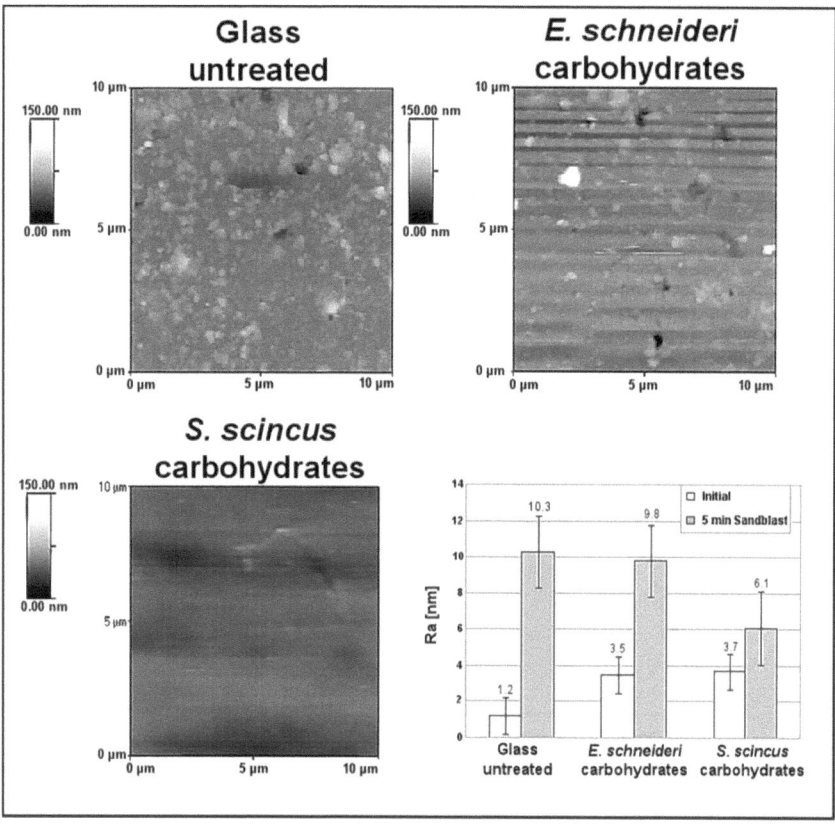

Figure 3.22 – Abrasion after sandblasting for 5 minutes from a height of 10 cm. Compared was untreated glass and silanised glycans of *S. scincus* and *E. schneideri*. The roughness (measured by AFM) indicates the abrasion that happened during treatment with sand.

Silanisation with synthetic carbohydrates

Silanisation was also performed with Sialyl-lactose and D-lactose (obtained from Prof. Dr. Hanisch; University of Cologne). AFM measurements of the topography confirmed a coating of glass with these reducing oligosaccharides, however no effect on reduction of the adhesion force could be observed (the attractive force was similar to untreated glass; data not shown).

3.4.3 Quantification of silanised glycans

A quantification of glycans, silanised on a glass coverslip was performed with concanavalin A coupled peroxidase as carbohydrate marker. Peroxidase reacts with TMB ready to use ELISA substrate resulting in a colour change to blue. The reaction was stopped after 10 minutes with sulphuric acid, shifting the colour to yellow. Through the absorption intensity, which was measured photometrically, the amount of linked lectin could be determined by a standard curve (fig. 3.23). The amount of linked lectin is giving evidence of immobilized carbohydrates on the coverslips (conA is specific for mannose).

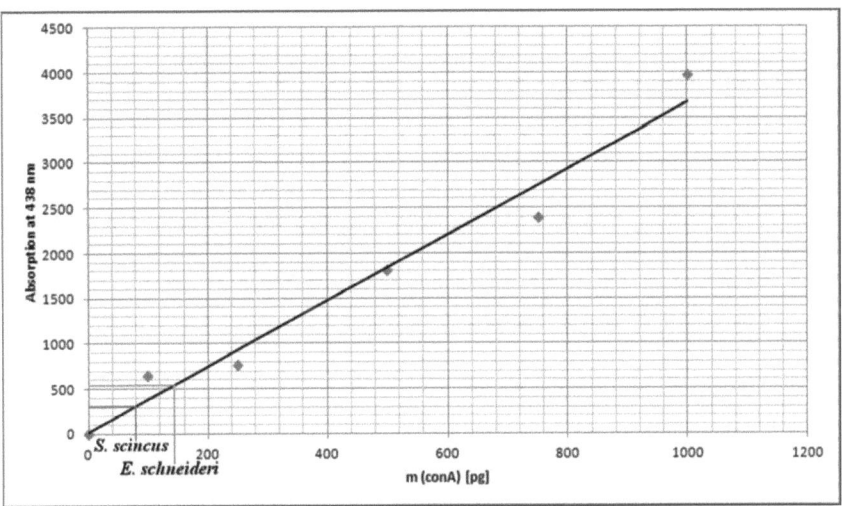

Figure 3.23 – Quantification of linked carbohydrates on a glass coverslip using concanavalin A coupled peroxidase as carbohydrate marker using a standard curve. Red: *Scincus scincus*; green: *Eumeces schneideri*.

The absorption at 438 nm of coverslips of S. scincus was $310.67 \pm SD = 38.68$ (n = 3) and of E. schneideri $532.67 \pm SD = 116.71$ (n = 3) after setting the zero point with coverslip controls, treated the same way as silanisation with reptilian glycans but without carbohydrates. The

Results

coverslip with glycans of S. scincus bound 80 pg and coverslips with glycans of E. schneideri 140 pg concanavalin A.

3.4.4 Monosaccharide classification

Results were generated by Prof. Hanisch, University of Cologne. Investigated were both prominent β-keratin bands at 17 kDa (see fig 3.17). The monosaccharide analysis of both proteins bands revealed mannitol. The analysis of the monosaccharides from deglycosylation of the skin exuviae (chapter 2.4.6) resulted in mannose (Man), glucose (Glc) and N-acetylglucosamine (GlcNAc). N-acetylgalactosamine (GalNAc) most likely was present, too; however it was covered by contaminations. Sialic acid (NeuAc) was detected at trace amounts only (table 3.1). Exuviae from *S. scincus* revealed a relative higher quantity of glycans than exuviae from *E. schneideri*. Monosaccharides found resemble the lectin specificity given in table 1.1.

Monosaccharide	Retention time [min]*	Characteristic fragments [m/z]	Presence
Fuc	9.6 (α)	204	-
Man	12.8 (α)	204	+
Gal	13.6/14.1(α, β)	204	-
Glc	14.5/14.9(α, β)	204	+
GalNAc	17.5(α, β)	173	(+)
GlcNAc	18.1(α)	173	+
NeuAc	22.7(α)	298, 420	(-)**

* Major signals of the pyranosidic α- or β- anomers are listed.

** Detectable at only trace amounts

Table 3.1 - Monosaccharide composition analysis by GC-MS of the TMS derivatives.

3.4.5 Glycan analysis

Preliminary glycan analysis resulted in no identifyable structures as investigated by Prof. Hanisch, University of Cologne. There are 3 glycosylated protein bands present in the skin lysate: 2 prominent bands are at ca. 17 kDa and 1 band at ca. 30 kDa (see chapter 3.3.4). The bands at 17 kDa do not contain N-linked glycans as found by PNGase F induced deglycosylation. β-elimination resulted in O-linked glycans of unknown structure. The structures were identified as glycans by MS-2 analysis, indicating oligosaccharids of 3-4 monomers with sialic acid as terminal saccharid. Glycans obtained from deglycosylation of complete exuviae (chapter 2.4.4) produced no stable glycans, because peeling hydrolyzed the structures prior to analysis.

3.4.6 Glycopeptide analysis

β-keratin protein bands at ca. 17 kDa were digested by trypsine and V8-protease by Prof. Hanisch, University of Cologne. A peptide mass analysis resulted in no consensus sequence; neither with the complete β-keratin gene nor with the fragment investigated by Saxe, 2008. Furthermore, no imminiumions could be verified, indicating the absence of N-linked glycans.

4. Discussion

4.1 Adaptations of *Scincus scincus* to a subterranean life

While many lizard species are able to burry themselves into lose sand (e.g., *Eumeces schneideri* and some species of the genus *Phrynosoma*), only a few lizard species have evolved the ability to perform real sandswimming in aeolian sand. Next to *Scincus scincus* and *Meroles anchietae* this behavior is for example shown by *Uma scoparia* and *Angolosaurus skoogi* (Arnold, 1995; Jayne and Daggy, 2000; Nance, 2007). So far, only little research has been made on this amazing ability; however it is known that this feature is acquired through some morphological adaptations (Hartmann, 1989) and that the epidermis shows a very low friction angle to sand together with a high abrasion resistance in comparison to technical surfaces (Baumgartner et al., 2007; Rechenberg et al., 2004; Rechenberg et al., 2009). The friction angle to sand of the sandswimming species *S. scincus*, *S. albifasciatus* and *M. anchietae* was also found to be much lower than that of chosen non-sandswimming reptilian species as investigated by Saxe, 2008 and in this study. The friction angles of the non-sandswimming species show a high variation, which may be caused as an adaptation to different habitats and life styles. *E. schneideri* shows the lowest value, which might has evolved due to its behaviour to burry itself into sand. Friction is necessary for movement and a low friction angle as the sandswimming species exhibit is surely of disadvantage in other habitats than sand deserts and a non-subterranean life style. The friction reduction of the epidermis in sandswimming species obviously evolved as an adaptation to the subterranean lifestyle, which enables an energy saving locomotion through sand.

Many physical effects are caused by the microstructure of the surface. Well known examples in the field of biomimetics are the superhydrophobicity of lotus plants (e.g., Barthlot et al., 2004) or drag reduction of shark skins (e.g., Dean and Bhushan, 2010). A specific microstructure can also have effects on the friction of a surface. One of the most prominent examples in nature is the gecko, where an increase of the relative contact surface through spatulae leads to adhesion through van der Waals forces (e.g., Autumn et al., 2002). The opposite, the reduction of frictional forces could be observed in snake lokomotion through "double-ridge" nanoscale microfibrillar geometry (Hazel et al., 1999). Furthermore, it could be shown that a micropattern (like knobs) can reduce friction through reduction of the relative contact area (roughness) (i.e., Varenberg and Gorb, 2009) that may lead to a decrease of adhesion through less van der Waals bonds. A surface ornamentation was also found in the sandfish and discussed to be casually involved in friction reduction.

Discussion

4.2 Microornamentaion of sandfish scales as cause for friction reduction

Though a micropattern can be utilized to reduce friction and may be important in some desert living species (i.e., Zhiwu et al., 2012), the serrated micro-structure, which is found on dorsal scales of the sandfish obviously has no effect on low friction and abrasion resistance due to the following reasons: (1) the comb-like structure is found on dorsal scales only, whereby the ventral scales are completely smooth, though they should be subjected to higher frictional forces through gravitation than the dorsal scales. However, ventral scales show the same low friction behavior and should have the same abrasion resistance as dorsal scales; (2) the closely related, non-sandswimming skink *Eumeces schneideri* as well as *Scincopus fasciatus* show a similar serrated dorsal structure like the sandfish (with a smooth ventral side in all three species), but these non-sandswimming skinks exhibit a much higher friction angle than the sandfish. Furthermore, these structures are also found on scales of other species like *Acontias percivali* (Schmied, 2007) and even in an aquatic snake, *Helicops modestus* (Rocha-Barbosa and Maraes e Silva, 2009); (3) the scales of another sandswimming species, *Meroles anchietae* (Lacertidae), which is not closer related to the sandfish, possess similar abilities of low friction but no specific micro-structure is found on dorsal or ventral scales; (4) resin replicas of dorsal serrations of the sandfish show no effect on abrasion resistance and friction angle in comparison to a smooth resin control. Though the serrations as cause for the skin properties of the sandfish are spaciously discussed (e.g., Rechenberg et al., 2004; Rechenberg et al., 2009) their influence for friction reduction can be excluded for these reasons. Therefore, other functions of the microstructures were investigated.

4.3 Possible functions of the microstructure on dorsal scales

Some other purposes of the serrated, comb-like structures on dorsal scales like radiation and moisture harvesting and the occurrence of triboelectric effects (described in Rechenberg et al., 2009) could not be verified by experiments, performed in this study. It is possible that the serrations are an evolutionary relict or serve as help in the moulting process to abrade old skin; it is suggested that the microstructur serves as a "zip-fastener" to hold the old and new skin together until the old skin is completely shed (Maderson, 1966), or that a surface strengthening is achieved by the microgrooves of the surface (Ruibal and Ernst, 1965). Though no purpose of the microornamentation was detected, it could be shown that they do not have an effect on friction. Therefore, the chemical composition of the scales was investigated to find out whether it has an impact on the skin's properties of the sandfish.

4.4 Chemical composition of the scales as cause of friction reduction

An influence of the material properties of moulted exuviae on friction was investigated by lysing the skin proteins and reconstituating them on a technical surface The same low adhesion force as the native scales exhibit could be verified (1) through dialysis of urea based skin lysate and reconstitution of the skin proteins on the dialysis membrane (Baumgartner et al., 2007) and (2) through the lysis of moulted exuviae with ammonium hydroxide and reconstituation of the solved proteins by adsorption on a glass surface in this study. These results give clear evidence that the lysed skin proteins enhance the anti-adhesive characteristics of the sandfish's epidermis and that the skin's properties are thus probably caused by the material composition. Furthermore, previous research revealed glycosylated β-keratins in the sandfish's exuviae (Baumgartner et al., 2007), of which the scales mainly constist of. Since the glycosylation of skin proteins may cause the properties of the sandfish epidermis, the chemical composition of the scale material was further investigated.

4.5 Analysis of the scale material
4.5.1 Biochemichal investigations

Moulted skin of the sandfish and the Berber skink was lysed and separated by an SDS-page. The lysate was separated mainly in two prominent bands at 17 kDa and one further band at 30 kDa, which are in the typical range of β-keratin proteins (Toni et al., 2007). Interestingly, the specific β-keratin antibody (which binds to the pre-corebox) reacted not only with the protein bands at 17 kDa, but also with a protein band at 30 kDa. Other proteins (like α-keratins) do not react with this antibody (Alibardi, personal information), so that there are at least two different types of β-keratins present in the investigated skink species.

In a further step, the skin proteins of the sandfish were investigated by western blot/ECL system with concanavalin A as carbohydrate marker. The results revealed that all three bands (both bands at 17 kDa and the band at 30 kDa), which were identified as β-keratins through the visualization with the specific antibody, are glycosylated. To find out the type of glycosylation, different enzymes or combinations were employed to deglycosylate these proteins. The two prominent bands at 17 kDa could not be deglycosylated by any enzyme or combination used. The reason why these protein bands were not affected by enzymatic treatment might be that an unusual type of glycosylation is present (e.g., O-mannosylation) or that the core structures of O-glycans are inaccessible for O-glycosidase. Though sialidase was used in this study to expose the core structure, this treatment is in some cases still insufficient and the additional use of hexosaminidase and β-galactosidase is required (Prozyme®, O-

Glycanse™, instruction manual). The other β-keratin at 30 kDa was successfully deglycosylated by the use of N-glycosidase. This finding is interesting, because skin treated with N-glycosidase lose low friction behavior (Baumgartner et al., 2007). Therefore, it is either possible that this protein leads to the characteristics of the sandfish skin or that N-glycans attached to this protein are necessary for the structural integrity of the skin.

Skin lysates of the Berber skink were used as control for the visualization of glycosylation. Surprisingly, this skink showed glycosylated β-keratins as well. As consequence, skin lysates of various sauropsidean species were investigated for glycosylation. Although, glycosylated proteins were found within the typical β-keratin range in virtually all species investigated, high quantities of glycosylated proteins could only be detected in the sandswimming species *S. scincus* and *M. anchietae*. Quantification confirmed that the amount of glycosylation is higher in the sandswimmers. In Baumgartner et al., 2007 (fig. 4) only the sandfish showed glycosylated β-keratins by comparison with *Pantherophis guttatus* (Colubridae) and *Crotapythus collaris* (Crotaphytidae). The reason why the latter species showed no glycosylation might be caused by following reasons: the quantity of protein was too little and PAS-reaction is not as sensitive as the ECL-system. Subsequently, these results confirm also a higher glycosylation in the sandfish. *Pantherophis guttatus* showed no glycosylation in this study as well when only 5 µg protein was applied in ECL-visualization.

It remains unclear which function glycans have that are attached on β-keratins of all sauropsidean species investigated and even found in mammalian hair (α-keratins). A possible reason might be that glycans serve not only to reduce friction in sandswimming species but also to increase abrasion resistance, which might be necessary in all species. Glycans may form a protection layer similar to an ice layer on asphalt, protecting beneath proteins and serve to reduce friction in sandswimming species. The assembly or chemical composition of these glycans may thus induce the physical properties, i.e., to protect from abrasion in all reptilian species and also to decrease friction in sandswimming species. However, the biochemical investigations clearly show that the glycosylation of β-keratins is higher in the sandswimmers *S. scincus* and *M. anchietae* in comparison to all other sauropsidean species examined. To prove this finding through a genetic foundation, the β-keratin coding gene was analysed.

4.5.2 DNA analysis

The isolation and sequencing of β-keratin coding genes was cumbersome. The most promising method is cDNA RACE is and sequenced β-keratin genes from other groups were obtained using this technique (e.g., Valle et al., 2005; Valle et al., 2007). The disadvantage of this method is that the animals need to be harmed to obtain living skin tissue that still produces mRNA. In this study, cDNA RACE could be performed after the natural death of a sandfish. However, no DNA was obtained, because the sandfish presumably was found and frozen too late and all mRNA was already hydrolized at this time. In further studies, if more β-keratin coding genes need to be examined, this method can be used on living species without harming the animals too much; it was observed that tails of own specimens were bitten off in combats and regrew in only a few weeks. Under the observation or help of a veterinary, the tip of the tail can be removed after local anaesthetization with e.g., lidocaine and the tissue immediately processed for mRNA isolation.

The ligation techniques performed (RNA hybridization and T4 RNA ligation) did not yield analyzable results as well. Amplified PCR products resulted in sequences that were not part of a β-keratin coding gene; either contaminations occurred, the primers used were not completely gene specific or one primer served as both forward and reverse primer.

The high abundance of similar sequences in both, β-keratin coding and non-coding regions resulted in the amplification of a complete gene. This was achieved by sequencing the start and terminus region with primers that are specific to β-keratin coding and non-coding parts of the genome. Furthermore, it was possible to amplify a promoter of the gene using this technique. Both genes of *Scincus scincus* and *Eumeces schneideri* show only little difference so that obviously homologue genes have been sequenced. However, the gene sequences clearly indicate some point mutations, which presumably occurred during speciation. The DNA sequences are short and no introns were detectable. This finding is in agreement with Alibardi et al., 2006 and Valle et al., 2005 where no introns could be identified in β-keratin coding genes of lizard species. In both sequences the corebox was included, which is conserved and thus highly homologue between all sauropsidian species that express β-keratins. The sequences themselves show no signal peptide consensus for posttranslational modification in the golgi apparatus (where usually O-glycans are linked to a protein); the prediction for translocation into the golgi apparatus for the β-keratin of *S. scincus* is only at 4.3% (PSORT II server; http://psort.hgc.jp/form2.html). However, this program prediction is based mainly on yeast cells. Tissue DNA of preserved specimen of *M. anchietae* was very unstable and the amplification of a β-keratin fragment with universal primers was difficult but possible. For complete gene amplification however, living specimen should be obtained. Though they are not commercial available, a permit for collection in Namibia can be obtained from Mr. Michael Griffin, Ministry of Environment and Tourism, Windhoek, Namibia.

Research permits are issued without further problems (Paul S. Freed, personal information; collector of specimen for McAllister et al., 2011). By comparison of the gene fragment of *M. anchietae* with *P. siculus*, a high similarity was found in both species as well. Similar to the sandfish, point mutations occurred during speciation and there were also no introns detectable in the fragment.

In-silico translated DNA to amino acid sequences show an evolutionary trend to exchange single amino acids to potential O-glycosylation sites through replacement with serine or threonine caused by point mutations. This is observed in sequences of both sandswimming species (*S. scincus* and *M. anchietae*) in direct comparison to related non-sandswimming species. The additional O-glycosylation sites in the sandfish reach also a higher score in the prediction for glycosylation, which indicates that these additional potent amino acids are presumably glycosylated. This result gives further evidence that a higher glycosylation is present in *Scincus scincus* in comparison to *Eumeces schneideri* and is thus giving molecular support that glycans attached to β-keratins are crucial to reduce friction. Though there is a higher abundance of O-glycosylation sites present in *M. anchietae* in comparison to *P. siculus* as well, the scores for these additional amino acids do not reach the threshold for glycosylation. It is possible that the low score is caused by the algorithm of the program used, which is based on human mucin glycans. In summary, the results acquired through biochemical and genetic examinations clearly indicate a higher glycosylation of the β-keratin proteins in sandswimming species. Therefore, the glycans were subjected for analysis to find out more about their possible function for friction reduction.

4.5.3 Glycan analysis

A glycopeptide analysis was performed to compare the mass of digested β-keratin proteins with the in-silico translated amino acid sequence to find out glycosylation sites. No peptide matching the translated β-keratin gene sequence was found, what could be caused by following reasons. There are more than 40 different β-keratins present in *Anolis carolinensis* (Valle et al., 2008). It can be assumed that an equal amount of β-keratins is present in *Scincus scincus* as well that makes it difficult to identify a single peptide sequence. Furthermore, there is a high abundance of glycosylation sites found on the protein sequence; the attached carbohydrates render the mass of the peptides, so that they cannot be identified anymore by mass spectroscopy without knowing the glycan assembly. MALDI-MS analysis to identify glycan structures of the most prominent β-keratins at 17 kDa resulted in data that could not be evaluated. However, carbohydrates were identified using a MALDI-2 analysis of single mass peaks (identified were sialic acid and hexoses). It can be assumed that glycans attached on the β-keratins consist of oligosaccharids of 3-4 monomers by the glycan masses. The experiments should be repeated and extended to the β-keratins present at 30 kDa. A monosaccharide analysis of β-keratin bands at 17 kDa revealed mannitol as the only carbohydrate detectable. This result needs to be repeated and verified to understand the possible role of this unusual carbohydrate for friction reduction. Carbohydrates that typically assemble glycans have been identified in samples, where the complete exuviae were deglycosylated by β-elimination. The analysis resulted in mannose (Man), N-acetylgalactose (GalNAc), N-acetylglucose (GlcNAc) and glucose (Glc). Sialic acid (NeuAc) was found in traces. Other residues like amino acids were not present. *S. scincus* showed a relative higher concentration of carbohydrates than *E. schneideri*, however there was no difference in types of monosaccharides between both species found. This indicates that the glycan assembly might be responsible for the anti-adhesive epidermis of *S. scincus*. Though the chemical analysis of glycans revealed only little data to evaluate, performed experiments in this study give clear evidence that glycans cause the friction reduction of the sandfish's epidermis.

4.6 Glycans as cause for friction reduction

Glycans were liberated and purified from the protein backbone by β-elimination and afterwards the released carbohydrates were covalently linked on a glass surface via silanisation. The adhesion force of this produced glycan array resembled the same low adhesion force of the native scales and the reconstituated proteins. Measurements of the friction angles of the glass covered with glycans showed the same low value as the native scales and a correlation was found between adhesion force and friction reduction. Remarkably, not only the adhesion force and friction of the sandfish glycans silanised on glass coverslips are similar to the native skin, but also the glycans of the Berber skink resemble the natural model. This clearly indicates that the glycan assembly renders the friction of the epidermis. Furthermore, a positive impact on abrasion resistance was found in silanised glycans of the sandfish. This may be caused by reduction of adhesive wear (less abrasion through less material interaction (friction)). At a longer term and harsher conditions however, the silanised glass lose abrasion resistance and low friction characteristics. It can be assumed that the high kinetic energy caused by sandblasting damaged the surface. Glass has a high elasticity modulus (50-80 GPa), whereby β-keratins on reptile skin are more elastic (gradient from 3-4 GPa in the sandboa *Gongylophis colubrinus*, Klein et al., 2010; 9-12 GPa for *Scincus scincus*, Baumgartner et al., 2007) that might be crucial to absorb kinetic energy and thus decrease abrasive wear. It is now necessary to immobilize glycans on a polymer surface that resembles the E-modulus as well as the elastic recovery of β-keratin layers to observe a long term impact on abrasion resistance.

The method used for silanisation was developed to link reducing carbohydrates for glycan arrays and a monosaccharide analysis of the samples verified carbohydrates that are typically present in glycan assemblies. Though the silanised glass coverslips of *Eumeces schneideri* reacted stronger with concanavalin A in relation to *Scincus scincus* this experiment is giving further clear evidence that indeed glycans were immobilised on the coverslips. Glycans of *Eumeces schneideri* bound more conA than *Scincus scincus*, indicating glycans that are more mannose-rich. The reason why there is a stronger reaction with conA of the native proteins in the ECL system of *S. scincus* is that its β-keratin proteins have a higher abundance of glycosylation sites and thus probably have a higher quantity of mannose containing glycans. The glycan arrays on glass immobilize the same quantity of glycans in the direct vicinity of each other and thus the glycan abundance in the proteins is irrelevant for this experiment. The results presented give therefore prove that carbohydrates are employed to reduce the friction of the epidermis of the sandfish. This finding leads to the conclusion that the glycans, and not the β-keratin proteins themselves are in direct contact with the environment and evidently render the epidermis of the sandfish anti-adhesive what directly reduces friction and has a positive impact on abrasion resistance.

The mechanism behind the low adhesion force is yet unknown. Capillary forces can be excluded in desert living species and AFM measurements were performed under dry conditions. Electrostatic forces cannot be completely excluded, but seem unlikely since both, the sample and the cantilever were attached to ground and thus should be electrically neutral. Furthermore, samples of *E. schneideri* were treated the same way as samples of *S. scincus* but show a higher attraction force, so that a charging of the samples through friction of the needle (attached to the cantilever) can be excluded. Presumably, the glycan composition of *S. scincus* reduces van der Waals forces; specific carbohydrates may prevent the formation of dipoles causing the van der Waals bonds.

4.7 Outlook

The biomimetic applications of an anti-adhesive surface, inspired by the sandfish, are extensive. In contrast to other surface coating techniques like e.g., the lotus-effect®, such a surface should be more stable to mechanical influences through a covalently bond of carbohydrates, is transparent and does not need the addition of water to clean. Therefore, it is possible to coat glass in windows, eyeglasses or monitors. The development of a high abrasion resistant surface might be achieved by linking the carbohydrates on a polymer that resembles the E-modulus and elastic recovery of the sandfish's β-keratin proteins.

It is now necessary to understand the mechanism how glycans are able to reduce adhesion forces, which possibly is caused by preventing the formation of van der Waals dipoles. Glycans usually are brachiated and two specific carbohydrates in the vicinity might preferably form a dipole with each other, and thus avert the emergence of a dipole with an approaching solid body. It is maybe even possible to calculate the optimal assembly of glycans to avoid the formation of dipoles with the help of a computer program. The effects of capillary and coulomb forces might be both completely excluded in further AFM measurements in an aqueous salt solution, so that only van der Waals forces should be measured.

In further studies, β-keratins that are responsible for the glycan based friction reduction have to be identified. Interestingly, it was demonstrated in Baumgartner et al., 2007 that exuviae of the sandfish, which was treated with PNGase F (which deglycosylates N-linked glycans) lose low friction behaviour. The results presented in this study clearly show that there are no N-linked glycans present in the two prominent β-keratin bands of the size of 17 kDa (examined by enzymatical deglycosylation, glycopeptides analysis and DNA sequence). Therefore it is possible that the β-keratins at 30 kDa promote the skin characteristics in combination with the prominent, highly glycosylated β-keratins at 17 kDa. The different β-keratin bands must be isolated by HPLC, or what might be easier due to the insolubleness of these proteins, by elution out of SDS-gels. Afterwards, they can be analysed (e.g., through MALDI-MS, monosaccharide analysis or deglycosylation and silanisation of the glycans on glass). As an additional experiement, the glycans of the moulted exuviae should be liberated and purified directly prior to MALDI-MS analysis. In this study, the deglycosylated glycans turned out to be not very stable after β-elimination and strong peeling reaction that occurred during storage and transport made an analysis of the glycan structure impossible.

Though the β-keratin coding gene of the sandfish is to date of completion of this study investigated for size, glycosylation and other properties in a bachelor thesis by cloning it into eukaryotic CHO cells, further β-keratin coding genes should be identified by performing cDNA RACE. Additional β-keratins might have other properties and are important to find out more about the glycosylation of these proteins.

References

Al-Sadoon, M. K., Al-Johany, A. M.-F. (1999). Food and Feeding habits of the sand fish lizard Scincus mitranus. *Saudi J. Bio. Sci.* **6**: 91-101.

Alibardi, L, Valle, LD, Toffolo, V, Toni M. (2006). Scale keratin in lizard epidermis reveals amino acid regions homologous with avian and mammalian epidermal proteins. *The Anatomical Record* **288A**: 734-752.

Arnold, E., Leviton, A. (1977). A revision of the lizard genus Scincus. *Bull. Br. Mus. nat. Hist. (Zool.)* **31**: 189-248.

Arnold, E. (1995). Identifying the effects of history on adaptation – origins of different sand-diving techniques in lizards. *J. Zool.* **235**: 351–388.

Autumn, K., Sitti, M., Liang, Y.A., Peattie, A.M., Hansen, W.R., Sponberg, S., Kenny, T.W., Fearing, R., Israelachvili, J.N., Full R.J. (2002). Evidence for van der Waals adhesion in gecko setae. *PNAS* **99**: 12252–12256.

Bauer, A.M. (1998). Lizards. In: Cogger, H.G., Zweifel, R.G. (Eds.), Encyclopedia of Reptiles and Amphibians, second ed. Academic Press, San Deigo: 126–173.

Baumgartner, W., Weiβ, P., Schindler, H. (1998). A Nonparametric Test for the General Two-Sample Problem. *Biometrics* **54**: 1129-1135.

Baumgartner W., Saxe F.P.M., Weth, A, Hajas, D., Sigumonrong, D., Emmerlich, J., Singheiser, M., Böhme, W., Schneider, J. (2007). The sandfish's skin: morphology, chemistry and reconstruction. *J. Bionic. Eng.* **4**: 1-9.

Baumgartner, W., Fidler, F., Weth, A., Habbecke, M., Jakob, P., Butenweg, C., Böhme, W. (2008). Investigating the Locomotion of the Sandfish in Desert sand using NMR Imaging. *PLoS ONE* **3**: e3309.

Barthlott, W., Cerman, Z., Stosch, A.K. (2004). Der Lotus-Effekt: Selbstreinigende Oberflächen und ihre Übertragung in die Technik. *Biologie in unserer Zeit* **34**: 290–296.

Beckmann, H. (2010). Darstellung von Kohlenhydrat-Mikroarrays unter Verwendung von Diels-Alder-Reaktionen mit inversem Elektronenbedarf. Dissertation Universität Konstanz. Logos Verlag, Berlin.

Berthé R.A. (2009). Surface structure and frictional properties of the Amazon tree boa Corallus hortulanus (Squamata, Boidae). *J. Comp. Physiol. A* **195**: 311-318.

Bradford, M.M. (1976). A rapid and sensitive method for the quantitation of microgram quantities of protein utilizing the principle of proteindye binding. *Anal. Biochem.* **72**: 248-254.

References

Carlson, D.M. Blackwell, C. (1968). Structures and Immunochemical Properties of Oligosaccharides Isolated from Pig Submaxillary Mucins. *J. Biol. Chem.* **243**: 616-626.

CC Technologies. (2001). Corrosion costs and preventive strategies in the United States. Publication No. FHWA-RD-01-156. From http://www.watermainbreakclock.com/docs/techbrief.pdf.

Comanns, P., Effertz, C., Hischen, F., Staudt, K., Böhme, W., Baumgartner, W. (2011). Moisture harvesting and water transport through specialized micro-structures on the integument of lizards. *Beilstein J. Nanotechnol.* **2**: 204-214.

Crowley, J.F., Goldstein, I.J., Arnap, J., Lonngren, J. (1984). Carbohydrate binding studies on the lectin from Datura stramonium seeds. *Arch. Biochem. Biophys.* **231**: 524–533.

Czichos, H. (1978). Tribology: A Systems Approach to the Science and Technology of Friction, Lubrication, and Wear. Tribology Series 1, New York, Elsevier Scientific Publishing Company.

Dean, B., Bhushan, B. (2010). Shark-skin surfaces for fluid-drag reduction in turbulent flow: a review. *Phil. Trans. R. Soc. A* **368**: 4775-4806.

de Boer, A.R., Hokke, C.H., Deelder, A.M., Wuhrer, M. (2007). General microarray technique for immobilization and screening of natural glycans. *Anal. Chem.* **79**: 8107-8113.

Goldstein, I.J., Hayes, C.E. (1978). The Lectins: Carbohydrate-Binding Proteins of Plants and Animals. *Advan. Carbohyd. Chem. Biochem.* **35**: 127–340.

Hansen, J.E., Lund, O., Nielsen, J.O., Brunak, S. (1996). O-glycbase: a Revised Database of O-glycosylated Proteins. *Nucleic Acids Res.* **24**: 248-252.

Hartmann, U.K. (1989). Beitrag zur Biologie des Apothekerskinks, *Scincus scincus*, Teil 2. *Herpetofauna* **11**: 12-24.

Hazel, J., Stone, M., Grace, M.S., Tsukruk, V.V. (1999). Nanoscale design of snake skin for reptation locomotions via friction anisotropy. *J. Biomech.* **32**: 477-484.

Holtzauer, M. (1997). Biochemische Labormethoden (3 ed.). Berlin: Springer Verlag.

Hazel J., Stone M., Grace M.S., Tsukruk V.V. (1999). Nanoscale design of snake skin for reptation locomotions via friction anisotropy. J Biomech **32**: 477-484.

References

Huang, Y., Mechref, Y., Milos, V. (2001). Microscale Nonreductive Release of O-Linked Glycans for Subsequent Analysis through MALDI Mass Spectrometry and Capillary Electrophoresis. *Anal. Chem.* **73**: 6063-6069.

Jones, M.H., Scott, D. Eds. (1983). Industrial Tribology: the practical aspects of friction, lubrication, and wear. New York, Elsevier Scientific Publishing Company.

Kalboussi, M., Aprea, G., Splendiani, A., Giovannotti, M., Caputo, V. (2006). Standard karyotypes of two populations of the *Scincus scincus* complex from Tunisia and Morocco (Reptilia: Scincidae). *Acta Herpetologica* **1**: 127-130

Klein M-C., Deuschle J., Gorb S.N. (2010). Material properties of the skin of the Kenyan sand boa Gongylophis colubrinus (Squamata, Boidae). *J. Comp. Physiol.* A **196**: 477-484.

Kohlmeyer, R. (2001). Agamen.de. Retrieved october 2, 2008, from http://www.agamen.de/reptilien/skinke/skinke.htm.

Jayne, B.C., Daggy M.W. (2000). The effects of temperature on the burial performance and axial motor patterns of the sand-swimming of the Mojave fringe-toed lizard Uma scoparia. *J. Exp. Biol.* **203**: 1241–1252.

Laemmli, U.K. (1970). Cleavage of structural proteins during the assembly of the head of bacteriophage T4. *Nature* **227**: 580-685.

Loyd, K.O., Burchell, J. Kudryashov, V., Yin, B.W.T., Taylor-Papadimitriou, J. (1996). Comparison of O-Linked Carbohydrate Chains in MUC-1 Mucin from Normal Breast Epithelial Cell Lines and Breast Carcinoma Cell Lines: DEMONSTRATION OF SIMPLER AND FEWER GLYCAN CHAINS IN TUMOR CELLS. *J. Biol. Chem.* **271**: 33325-33334.

Maderson, P.F.A. (1966). Histological changes in the epidermis of the Tokay (Gekko gecko) during the sloughing cycle. *J. Morphol.* **116**: 39–50.

Maladen, R.D, Ding, Y., Li C., Goldman D.I. (2009). Undulatory swimming in sand: subsurface locomotion of the sandfish lizard. *Science* **325**: 314-318.

McAllister, C.T., Bursey, C.R., Freed, P.S. (2011). Endoparasites (Cestoidea, Nematoda, Pentastomida) of Reptiles (Sauria, Ophidia) from the Republic of Namibia. *Comparative Parasitology* **78**: 140-151.

Merkle, R.K., Poppe, I. (1994). Carbohydrate composition analysis of glycoconjugates by gas-liquid chromatography-mass spectrometry. *Methods Enzymol.* **230**: 1-15.

References

Montreuil, J. (**1982**) Glycoproteins. In: A. Neuberger and L.L.M. Van Deenen (Eds), Comprehensive Biochemistry. Elsevier, Amsterdam: 1-188.

Morelle, W., Guyetant, R., Strecker, G. (1998). Structural analysis of oligosaccharide-alditols released by reductive beta-elimination from oviducal mucins of *Rana dalmatina*. *Carbohydr. Res.* **306**: 435-443.

Nance, H.A. (2007). Cranial osteology of the African gerrhosaurid Angolosaurus skoogi (Squamata; Gerrrhosauridae). *Afr. J. Herpetol.* **56**: 39–75.

Neuhäuser, M., Baumgartner, W., Weiβ, P. (2002). The Baumgartner-Weiβ-Schindler Test in the Presence of Ties. *Biometrics* **58**: 250-251.

Patsos, G., Andre, S., Roeckel, N., Gromes, R., Gebert, J., Kopitz, J., Gabius, H.J. (2009). Compensation of loss of protein function in microsatelliteunstable colon cancer cells (HCT116): A gene-dependent effect on the cell surface glycan profile. *Glycobiology* **19**: 726-734.

Prozyme® Glyco®, O-Glycanse™. Instruction manual. Retreived 2012, from http://www.prozyme.com/documents/gk80090.pdf.

Rechenberg, I., El Khyari, A.R. (2004). Reibung und Verschleiβ am Sandfisch der Sahara. Retrieved 2012 from http://www.bionik.tu-berlin.de/institut/festo04.pdf.

Rechenberg, I., Zwanzig, M., Zimmermann, S., El Khyari, A.R. (2009). Tribologie im Wustensand. Sandfisch, Sandboa und Sandschleiche als Vorbild für die Reibungs- und Verschleisminderung. Schlussbericht, BMBF-Forderkennzeichen 0311967A: Laufzeit 01.02.06-31.03.09. Retrieved 2012 from http://www.bionik.tu-berlin.de/institut/TriboDueSa.pdf.

Rocha-Barbosa, O., Maraes e Silva, R.B. (2009). Analysis of the microstructure of Xenodontinae snake scales associated with different habitat occupation strategies. *Braz. J. Biol.* **69**: 919-923.

Ruibal R., Ernst V. (1965). The structure of the digital setae of lizards. *J. Morphol.* **117**: 271–294.

Rybicki, E., von Wechmar, M. (1982). Enzyme-assisted immune detection of plant virus proteins electroblotted onto nitrocellulose paper. *J. Virol. Methods* **5**: 267-278.

Sägerstrom, C.G., Sive H.L. (1996). RNA blot analysis. In: A Laboratory Guide to RNA: Isolation, Analysis, and Synthesis.Edited by: P.A. Krieg. Wiley-Liss Inc., New York, NY.

Sambrook, J. (2001a). Molecular Cloning: A laboratory manual. Volume 1, chapter 7.4. Cold Spring Harbor Laboratory Press, New York.

References

Sambrook, J. (2001b). Molecular Cloning: A laboratory manual. Volume 2, chapter 8.46-8.61. Cold Spring Harbor Laboratory Press, New York.

Sawyer, R., Knapp, L. (2003). Avian skin development and the evolutionary origin of feathers. *J. Exp. Zool. B* **298**: 57-72.

Saxe, F.P.M. (2008). Functional and morphological investigation of the epidermis of the sandfish (Reptilia: Squamata: Scincus scincus). Diploma Thesis, RWTH Aachen University.

Schmied, H. (2007). Rasterelektronenmikroskopische Analyse der Oberflächenstruktur der Schuppen verschiedener Arten von Glattechsen (Scincidae) der Gattungen *Scincus, Scincopus, Eumeces* und *Acontias*. Blockpraktikum, Zoological Research Museum Alexander Koenig.

Shibuya, N., Goldstein, I.J., Broeknaert, W.F., Nsimba-Lubaki, M., Peeters, B., Peumans, W.J. (1987). The elderberry (Sambucus nigra L.) bark lectin recognizes the Neu5Ac(alpha 2-6)Gal/GalNAc sequence. J. Biol. Chem. **262**: 1596–1601.

Shibuya, N., Goldstein, I.J., Van Damme, E.J.M., Peumans, W.J. (1988). Binding properties of a mannose-specific lectin from the snowdrop (*Galanthus nivalis*) bulb. J. Biol. Chem. **263**: 728.

Schmitz, A, Mausfeld, P, Embert, D. (2004). Molecular studies on the genus Eumeces. Wiegmann, 1834: Phylogenetic relationship and taxonomic implications. *Hamadryad* **28**: 73-89.

Song, X., Lasanajak, Y., Xia, B., Smith, D.F., Cummings, R.D. (2009). Fluorescent glycosylamides produced by microscale derivatization of free glycans for natural glycan microarrays. *ACS Chem Biol.* **4**: 741-750.

Taylor, M.E., Drickamer K. (2006). Introduction to glycobiology, 2nd Edition. Oxford University Press, USA.

Toni, M., Valle, L.D., Alibardi, L. (2007). Hard (Beta-) keratins in the epidermis of reptiles: composition, sequence, and molecular organization. *Journal of Proteome Research* **6**: 1792–1805.

Staudt, K., Saxe, F.P.M., Schmied, H., Böhme W., Baumgartner W. (2011). Sandfish inspires engineering. Bioinspiration, Biomimetics and Bioreplication, edited by Raul J. Martin-Palma, Akhlesh Lakhtakia, Proceedings of SPIE Vol. 7975 (SPIE, Bellingham, WA, 2011) 79751B.

Valle, L.D., Toffolo, V., Belvedere, P., Alibardi, L. (2005). Isolation of a mRNA encoding a glycine-proline-rich b-keratin expressed in the regenerating epidermis of lizard. *Dev Dyn* **234**: 934–947.

Valle, L.D., Nardi, A., Belvedere, P.T., Alibardi, L. (2007). B-keratins of differentiating epidermis of snake comprise glycine-proline-serine-rich proteins with an avian-like gene organization. *Dev. Dyn.* **236**: 1939-1953.

References

Valle, L.D., Nardi, A., Bonazza, G., Zuccal, C., Emera, D., Alibardi, L. (2008). Forty keratin-associated β-proteins (β-keratins) form the hard layers of scales, claws, and adhesive pads in the green anole lizard, Anolis carolinensis. *J. Exp. Zool. B* **314**: 11-32.

Varenberg, M., Gorb, S.N. (2009). Hexagonal Surface Micropattern for Dry and Wet Friction. *Advanced Materials* **21**: 483-486.

Wang, W.C., Cummings, R.D. (1988). The immobilized leukoagglutinin from the seeds of Maackia amurensis binds with high affinity to complextype Asn-linked oligosaccharides containing terminal sialic acid-linked alpha-2,3 to penultimate galactose residues. *J. Biol Chem* **263**: 4576–4585.

Wong, C., Burgess, J. (2002). Modifying the Surface Chemistry of Silica Nano-Shells for Immunoassays. *Journal of Young Investigators*: 6.

Xia, B., Kawar, Z.S., Ju, T., Alvarez, R.A., Sachdev, G.P., Cummings, R.D. (2005). Versatile fluorescent derivatization of glycans for glycomic analysis. *Nat Methods.* **2**: 845-850.

Young, J.D., Tsuchiya, D., Sandlin, D.E., Holroyde, M.J. (1979). Enzymatic O-glycosylation of synthetic peptides from sequences in basic myelin protein. *Biochemistry* **18**: 4444-4448.

Zhang, X-H., Chiang, V.L. (1996). Single-Stranded DNA Ligation by T4 RNA Ligase for PCR Cloning of 5'-Noncoding Fragments and Coding Sequence of a Specific Gene. *Nucl. Acids Res.* **24**: 990-991.

Zhiwu, H., Junqiu, Z., Chao, G., Li, W., Ren, L. (2012). Erosion resistance of bionic surfaces inspired from desert scorpions. *Langmuir* **28**: 2914-2921.

Acknowledgements

I am grateful to Prof. Dr. Werner Baumgartner for his inspirational suggestions and his helpful hand contributing much in the progress of this thesis and Prof. Dr. Wolfgang Böhme for having an open ear and solutions for nearly every problem.

I thank Prof. Dr. Hanisch (University of Cologne) for the cooperation to identify glycans, Prof. Dr. Alibardi (University of Bologna) for providing β-keratin specific antibodies, Prof. Dr. Göllner (RWTH Aachen University) for supplying materials and methods to perform cDNA RACE and the staff of the Fraunhofer Institute (IME), especially Raphael Souer for the fatigueless effort in sequencing DNA samples.

Very much appreciated is the work of Dipl.-Biol. Friederike Saxe and Dipl.-Biol. Heiko Schmied. A large part of my investigations are built upon their results.

Furthermore, I thank our technical staff, Agnes Weth for supporting this thesis during the first month and the last year and Dipl.-Biol. Christine Böhme for having a clear mind and soul.

I wish to acknowledge all my colleagues for providing a friendly and creative atmosphere, mostly Florian Hischen, Dipl.-Biol. Michael Bennemann, Dipl.-Biol. Phillip Comanns, Dr. Ralph Meisterfeld, Dr. Lars Dolge, Dr. Ingo Scholz and Dipl.-Biol. Thomas Erlanghagen.

I am grateful to my parents and friends, who supported me morally during difficult times.

Finally, the Deutsche Forschungsgemeinschaft (DFG) is acknowledged for financial support within the PhD-program (Graduiertenkolleg GRK1572). I also wish to thank all the members of this Graduiertenkolleg for their suggestions and presentations to give a deeper insight into the world of bionics.

Bibliographic notes

Within the investigations made in this thesis, some research data have been published or are to the date of completion in review for publishing to disseminate the most important scientific results presented to international viewers. This involves results regarding surface characteristics, DNA, protein and glycan investigations. The references for published data are:

Staudt, K., Saxe, F.P.M., Schmied, H., Böhme, W., Baumgartner, W. (2011). Sandfish inspires engineering. *Bioinspiration, Biomimetics, and Bioreplication*, edited by Raúl J. Martín-Palma, Akhlesh Lakhtakia, Proceedings of SPIE Vol. 7975 (SPIE, Bellingham, WA, 2011) 79751B.

Staudt, K., Saxe, F.P.M., Schmied, H., Souer, R., Böhme, W., Baumgartner, W. (2012). Comparative investigations of the sandfish's β-keratin (Reptilia: Scincidae: *Scincus scincus*) – Part 1: Surface and molecular examinations. *J. Biomim. Biomater. Tissue Eng.* Accepted.

Staudt, K., Böhme, W., Baumgartner, W. (2012). Comparative investigations of the sandfish's β-keratin (Reptilia: Scincidae: *Scincus scincus*) – Part 2: Glycan based friction reduction *J. Biomim. Biomater. Tissue Eng.* Accepted.

Appendix

Friction angles

Friction angles of native scales

Native Scales		*Scincopus fasciatus*	*Eumeces schneideri*	*Podarcis siculus*	*Pantherophis guttatus*	*Meroles anchietae*	*Scincus scincus*	*Scincus albifasciatus*
Dorsal		31	25	26	27	24	21	19
		32,5	26	27	28	25	21	21
		30	25	26	29	25	21	18
		30	24	28	28	23	20	17
		29	26	27	28	22	20	17
		32	26	32	27	23	21	19
		30	27	30	30	25	21	20
		31	26	30	28	24	20	21
		31	25	30	29	25	21	23
		31	26,5	31	31	25	20	23
Ventral		33	25	29	28	19	23	18
		33	26	26	26	20	24	20
		33	27	27	27	19	23	18
		34	26	28	29	20	22	18
		33	27	27	28	18	22	21
		35	26	27	32	17	19	20
		34	27	26	31	19	18	20
		34	28	27	29	19	19	19
		34	28	27	30	19	18	19
		34	27	28	29	19	19	19
Mean		32.2	26.2	28.0	28.7	21.5	20.7	19.5
SD		1.74	1.04	1.79	1.53	2.84	1.63	1.70
Remark		*	*			*	*	*

* Investigated by Saxe, 2008

Friction angles of glycans silanised on glass coverslips

	Glass untreated	*Eumeces schneideri* glycans	*Scincus scincus* glycans
Sample 1	26	26	21
	27	25	22
	28	24	19
	26	26	20
	27	25	20
	27	25	19
	28	24	22
	29	27	22
	28	26	21
	27	28	20
Sample 2	27	25	18
	28	24	18
	26	26	19
	29	27	18
	27	28	19
	28	27	20
	27	27	18
	26	24	21
	28	25	20
	28	27	19
Mean	*27.4*	*25.8*	*19.8*
SD	*0.93*	*1.32*	*1.36*

Appendix

Adhesion forces

	F [nN]	SD	n
Glass untreated	27	5	7
S. scincus native scales	2.1	1.4	5
S. scincus adsorbed proteins	2.5	1.4	4
S. scincus silanised glycans	2.4	1.6	5
E. schneideri native scales	16.5	3.2	4
E. schneideri adsorbed proteins	15.2	3.4	4
E. schneideri silanised glycans	17.1	4.1	4

Abrasion resistance

	Roughness [nm]	SD	n
	Initial		
Glass	1.2	0.9	6
E. schneideri	3.5	1.0	6
S. scincus	3.7	1.1	6
	5 min sandblast		
Glass	10.3	2.0	6
E. schneideri	9.8	1.9	6
S. scincus	6.1	2.0	6

Glycosylation intensities

Species	Grey value	Percentage in relation to background
S. scincus	49.1	0.75
M. anchietae	175.7	0.10
C. suchus	179.2	0.09
E. schneideri	192.4	0.02
G. gallus	194	0.01
P. guttatus	195.4	0.00
P. siculus	195.9	0.00
G. gecko	195.9	0.00
T. alba	195.9	0.00

Quantification of silanised glycans

Absorption (438 nm)	*Scincus scincus*	*Eumeces schneideri*
Sample 1	266	449
Sample 2	333	666
Sample 3	333	483
Mean	310.67	532.67
SD	38.68	116.71

Gene Sequences and protein translations

Primers

b-ker uni fw: 5' – CTG TGG TCC ATC CTG CGC TGT
b-ker uni rv: 5' – GCA CAT GGA GTG TTG CCT CCA AC Rev!
konr-MP fw: 5' – AACATTCCAGCAATCTTGGCA
konr-MP rv: 5' – AGCCAGTGCGCCAGCACCATA Rev!
SF_Start fw: 5' – ATGGCTGCTTGTGGTACATCTTGC
ES_Start fw: 5' – ATGGCTGCTTGTGGTCCATCTTGC
SF_Stop rv: 5' – CTAATAACAAATGTTACCGCGACGCC Rev!
Promoter: 5' – CATTCAGAGATAAATAAGGCTGTTCTTCAAGAAAGGTTAAAAACAAGGGA

Fragment Saxe, 2008 (*S. scincus*)

CTGTGGTCCATCCTGCGCTGTCCCATCCATCGCCTCCAGGCCTGTCGTTGGCTTTGGATCAGCAGGGTCGGGCTT
GGGCTATGGTCTCGGCTACGGAGGTTTTGGCTTAGGCTATGGCTATGGTGCTGGCGCACTGGCTAACATTCCAG
CAATCTTGGCACCCCGGCTGGAGTCATCCCTTCATGCATCAACCAGATCCCACCAGCAGAGGTTGTGATCCAGCC
ACCCTCTTCGATCGTGACCATCCCAGGACCCATCCTCTCTGCTAGCTGCGAGCCTGTTTCTGTTGGAGGCAACAC
TCCATGTGC

Start Sequenz (*S. scincus*)

AACATTCCAGCAATCTTGGCAAAAACGAAGAAGTGACCAT**CATTCAGAGATAAATAAGGCTGTTCTTCAAGAAAG
GTTAAAAACAAGGGA**TAAAATGATAAAGATGAAGAATTAAGGAACTGGGGTTTTAATATCTCTAGAAGTATGACA
TATGGTTTGAAGGAGGATGGCTTGAAGGATGGAGAACAGCAGGAGGAGATTGAGTGGGAAATAGGGCAACAGTTG
CAATAGCTGAAGAACAGCGTAGCTAAAACATCTCCTGCACTTCCTTTTGAATCTTGTACATTGTATCATTCTGTC
CCTTGATGTTTCTTTCAGACTAAACTGCATCAAACC****START****ATGGCTGCTTGTGGTACATCTTGCG
CTGTCCCATCCTGCGCTTCTAGCCCTGTTGTTGGCTTTGGATCGGCAGGTCTTGGAGGCTATGGAGGTTTTGGAG
GTCTCAGCTATGGCTATGGAGGTCTTGGTTATGGATATGGTGCTGGCGCACTGGCT

Appendix

Stop Sequence (*S. scincus*)

AACATTCCAGCAATCTTGGCACCCCGGCTGGAGTCATCCCTTCATGCATCAACCAGATCCCACCAGCAGAGGTTG
TGATCCAGCCACCCTCTTCGATCGTGACCATCCCAGGGCCCATCCTCTCTGCTAGCTGCGAGCCTGTTTCTGTTG
GAGGCAACACTCCCTGTGCTGTTGGTGGTTCTGGGATTGCCAGAGGCAGCTTCTTTGGGGGGTTGGGTTCTGGGG
CCTTCGGTTATGGGTATGGCCTGAGAGGGCGTGGCTACCTTGGAAGTTCTCTCCAGGGGCGTCGCGGTAACATTT
GTTATTAG***STOP***ACATGTAAACCTGAGCCTAGCAATCTAACCAGGAATGGATAAAATAATACACAGCA
ATCGCTGATTTTCTTTGAGATGGAGTTAGAAAAGCTAAACTTCTGGCACCATTATTTAAGTCCACAACTCTCCTT
TTCCCCTTTCTTTTCCCAAAGTCTCAAAACAAAGAATAAGTTCAGCTCTGTATTTTCTATTGTTTTGTCTTGTTC
TTCAACTGCTTCTTACACAATTGTTCATTCAACTCTTTTTCCTCATTTTTATTAAAATGTATGCTTCATTCTAA
TATGAGTTTGTGTGATTTTAATATGAGTTTTGGTATAGTGTTATATAGTCACTCCAATATGATTTTTTGTGT
AGTTCCTATTTCATAAAATGTGGCTGGGGTGCCAAGATTGCTGGAATGTT

Hybrid sequenze (merged Start and Stop sequenz, *S. scincus*)

CATTCAGAGATAAATAAGGCTGTTCTTCAAGAAAGGTTAAAAACAAGGGATAAAATGATAAAGATGAAGAATTAA
GGAACTGGGGTTTTAATATCTCTAGAAGTATGACATATGGTTTGAAGGAGGATGGCTTGAAGGATGGAGAACAGC
AGGAGGAGATTGAGTGGGAAATAGGGCAACAGTTGCAATAGCTGAAGAACAGCGTAGCTAAAACATCTCCTGCAC
TTCCTTTTGAATCTTGTACATTGTATCATTCTGTCCCTTGATGTTTCTTTCAGACTAAACTGCATCAAACC***
START***ATGGCTGCTTGTGGTACATGTTGCGCTGTCCCATCCTGCGCTTCTAGCCCTGTTGTTGGCTTTGGAT
CGGCAGGTCTTGGAGGCTATGGAGGTTTTGGAGGTCTCAGCTATGGCTATGGAGGTCTTGGTTATGGATATGGTG
CTGGCGCACTGGCTGAACATTCCAGCAATCTTGGCACCCCGGCTGGAGTCATCCCTTCATGCATCAACCAGATCC
ACCAGCAGAGGTTGTGATCCAGCCACCCTCTTCGATCGTGACCATCCCAGGGCCCATCCTCTCTGCTAGCTGCG
AGCCTGTTTCTGTTGGAGGCAACACTCCCTGTGCTGTTGGTGGTTCTGGGATTGCCAGAGGCAGCTTCTTTGGGG
GGTTGGGTTCTGGGGCCTTCGGTTATGGGTATGGCCTGAGAGGGCGTGGCTACCTTGGAAGTTCTCTCCAGGGGC
GTCGCGGTAACATTTGTTATTAG***STOP***

MAACGTSCAVPSCASSPVVGFGSAGLGGYGGFGGLSYGYGGLGYGYGAGALAEHSS
NLGTPAGVIPSCINQIPPAEVVIQPPSSIVTIPGPILSASCEPVSVGGNTPCAVGGSGIAR
GSFFGGLGSGAFGYGYGLRGRGYLGSSLQGRRGNICY*

Compete Gene (*S. scincus*)

START***ATGGCTGCTTGTGGTACATGTTGCAGCGTCCCATCCTGCGCTTCCACCCCGTTGTAGGCTTTGGAT
CAGGTGGTATTGGAGGGGGCTACGGAGGTCTTGGCCTGGGCTATGGATTTGGAGGCCTTGGTTATGGATATGGTG
CCGGAGGCCTGGCTGAAACTTCTGGAAATCTCGGCACCCTGGCTGGAGTCATCCCTTCCTGCATCAACCAGATCC
ACCAGCAGAGGTTGTCATCCAGCCACCTGCCTCCATTGTGACCATCCCTGGGCCCATCCTCTCTGCTAGCTGCG
AGCCCGTTGCTGTTGGAGGCAACACTCCCTGTGCAGTTGGCGGTTCTGGGATTGTAGGATCAGGTTTATTGGGAT
CAGGCCTCTATGGGGGTTTCGGCTATGGGGGCTTGGGTTACGGAGGCCTGGGTTATGGTTATGGCCTGAGGAGGG
GTGGATTCTTTGGAAGAAGGTCTCTCCTGAGGCGTCGCGGTAACATTTGTTATTAG***STOP***

Appendix

MAACGTSCTVPSCASTPVVGFGSGGIGGGYGGLGLGYGFGGLGYGYGAGGLAETSG
NLGTLAGVIPSCINQIPPAEVVIQPPASIVTIPGPILSASCEPVAVGGNTPCAVGGSGIVG
SGLLGSGLYGGFGYGGLGYGGLGYGYGLRRGGFFGRRSLLRRRGNICY*

Start Sequence (*E. schneideri*)

```
CATTCAGAGATAAATAAGGCTGTTCTTCAAGAAAGGTTAAAAACAAGGGATAACATGATAAAGATAAAGAATTAA
GGAATTGGGGTTTAATATCTCTAGAAGTATGACCTAGGGTTTGAAGGAGGATGTCTTGAAGGATGGAGAACAGCA
GGAGGAGGTTGAGTGGGAAACAGGGCAACAGTTGCAATAGCTGAAGAACAATGTAGATAAAACATCTCCTGCACT
TCCTTTTGAATTTCGTACATTGTGTCATTCTGTCCCTTGATGCTTTCTTTCAGATTAAACTCCATCAAACC***S
TART***ATGGCTGCTTGTGGTCCATCTTGTGCTGTCCCATCCTGTGCTTCTAGCCCTGTTGTAGGCTTTGGATC
GGCAGGTCTTGGAGGTCTTGGATGGGGCTATGGAGGTCTTGGAGGTCTCAGCTATGGATATGGAGGTCTCGGTTA
TGGTGTCGGCGGCCTGGCTGAAACTTCTGGCAATCTTGGCACCCTAGCTGGAGTCATCCCTTCCTGCATAAACCA
GATCCCACCAGCAGAGGTTGTCATCCAGCCACCCGCCTCCATTGTGACCATCCCAGGACCCATCCTNTCTGCTAG
CTGCGAGCCCGTTTCTGTTGGAGGCAACACTCCATGTGC
```

Stop Sequence (*Eumeces schneideri*)

```
AACATTCCAGCAATCTTGGCACCCCGGCTGGAGTCATCCCTTCATGCATCAACCAGATCCCACCAGCAGAGGTTG
TGATCCAGCCACCCTCTTCGATCGTGACCATCCCAGGACCCATCCTCTCTGCTAGCTGCGAGCCTGTTTCTGTTG
GAGGCAACACTCCCTGTGCTGTTGGTGGTTCTGGGATTGCCAGAGGCAGCTTCTTTGGGGGGTTGGGTTCTGGGG
CCTTCGGTTATGGGTATGGCCTGAGAGGGCGTGGCTATCTTGGAAGTTCTCTCCAGGGGCGTCGCGGTAACATTT
GTTATTAG***STOP***ACATGTGAACCTGAGCCTAGCAATCTAACCAGGAATGGATAAAATAATACACAGCAA
TCGCTGATTTTCTTTGAGATGGAGTTAGAAAAGCTAAACTTCTGGCACCATTATTTAAGTCCACAACTCTCCTTT
TCCCCTTTCTTTTCCCAAAGTCTCAAAACAAAGAATAAGTTCAGCTCTGTATTTTCTATTGTTTTGTCTTGTTCT
TCAACTGCTTCTTACACAATTGTTCATTCAACTCTTTTTTCCTCATTTTTATTAAAATGTATGCTTCATTCTAAT
ATGAGTTTGTGTGATTTTAATATGAGTTTTTGATATAGTCTAATCTAGTCACTCCAATATGATTTTTTGTGTA
GTTCCTATTTCATAAAATATGGCTGGGTGCCAAGATTGCTGGAATGTT
```

Hybrid sequence (merged Start and Stop sequenz, *E. schneideri*)

```
CATTCAGAGATAAATAAGGCTGTTCTTCAAGAAAGGTTAAAAACAAGGGATAACATGATAAAGATAAAGAATTAA
GGAATTGGGGTTTAATATCTCTAGAAGTATGACCTAGGGTTTGAAGGAGGATGTCTTGAAGGATGGAGAACAGCA
GGAGGAGGTTGAGTGGGAAACAGGGCAACAGTTGCAATAGCTGAAGAACAATGTAGATAAAACATCTCCTGCACT
TCCTTTTGAATTTCGTACATTGTGTCATTCTGTCCCTTGATGCTTTCTTTCAGATTAAACTCCATCAAACC***S
TART***ATGGCTGCTTGTGGTCCATCTTGTGCTGTCCCATCCTGTGCTTCTAGCCCTGTTGTAGGCTTTGGATC
GGCAGGTCTTGGAGGTCTTGGATGGGGCTATGGAGGTCTTGGAGGTCTCAGCTATGGATATGGAGGTCTCGGTTA
TGGTGTCGGCGGCCTGGCTGAACATTCCAGCAATCTTGGCACCCCGGCTGGAGTCATCCCTTCATGCATCAACCA
GATCCCACCAGCAGAGGTTGTGATCCAGCCACCCTCTTCGATCGTGACCATCCCAGGACCCATCCTCTCTGCTAG
CTGCGAGCCTGTTTCTGTTGGAGGCAACACTCCCTGTGCTGTTGGTGGTTCTGGGATTGCCAGAGGCAGCTTCTT
TGGGGGGTTGGGTTCTGGGGCCTTCGGTTATGGGTATGGCCTGAGAGGGCGTGGCTATCTTGGAAGTTCTCTCCA
GGGGCGTCGCGGTAACATTTGTTATTAG***STOP***
```

Appendix

MAACGPSCAVPSCASSPVVGFGSAGLGGLGWGYGGLGGLSYGYGGLGYGVGGLAE
HSSNLGTPAGVIPSCINQIPPAEVVIQPPSSIVTIPGPILSASCEPVSVGGNTPCAVGGSGI
ARGSFFGGLGSGAFGYGYGLRGRGYLGSSLQGRRGNICY*

Complete Gene (*E. schneideri*)

START***ATGGCTGCTTGTGGTCCATCTTGCACCGTCCCATCCTGCGCTTCCAGCCCTGTTGGAGGCTTTGGAT
CAGGTGGGATTGGAGGAGGCTACGGAGGTCTCGGTTATGGATTTGGAGGCCTTGGTTATGGATATGGTGCTGGAG
GCCTGGCCGAAACTTCTGGAGATCTCGGCACCCTGGCTGGAGTCATCCCTTCCTGCATCAACCAGATCCCACCAG
CAGAGGTTGTCATCCAGCCACCCGCCTCCATTGTGACCATCCCTGGGCCCATCCTCTTTGCTAGCTGCGAGCCCG
TTGCTGTTGGAGGCATCACTCCCTGCGCTGCTGGTGGTTCTGGGGTTACAGGATCAGGTTTATTGGGATCTGGCC
TCTATGGGGGTTTCGGCTATGGGGGCTTGGGTTACGGAGGCCTGGGTTATGGTTATGGCCTGAGGAGGGGTGGAT
TCTTTGGAAGAAGGTCTCTCCTGAGGCGTCGCGGTAACATTTGTTATTAG***STOP***

MAACGPSCTVPSCASSPVGGFGSGGIGGGYGGLGYGFGGLGYGYGAGGLAETSGDL
GTLAGVIPSCINQIPPAEVVIQPPASIVTIPGPILFASCEPVAVGGITPCAAGGSGVTGSG
LLGSGLYGGFGYGGLGYGGLGYGYGLRRGGFFGRRSLLRRRGNICY*

Fragment (*M. anchietae*)

CTGTGGTCCATCCTGCGCTGTCCCATCCTGTGCTTCAACCCCCACCGTTGGATTTGGATCAGCAGGTGGTCTTGG
CTATGGAGTTCTTGGTAGAGGGGAGGTCTCGGCTATGGATACGGAGGTCTCGGCTATGGTTTTGGAGGTGCTGA
AAGAGCAACCAACCTTGGAATCCTGGCAGGAGTTGTCCCATCATGTGTCAACCAGATCCCACCAGCAGAGGTTGT
GATCCAGCCACCCCCCACCGTCCTGACCATCCCAGGGCCCATCCTCTCTGCCAGCTGTGAGCCAGTGGCTGTTGG
AGGCAACACTCCATGTGC

CGPSCAVPSCASTPTVGFGSAGGLGYGVLGRGGGLGYGYGGLGYGFGGAERATNLG
ILAGVVPSCVNQIPPAEVVIQPPPTVLTIPGPILSASCEPVAVGGNTPC

Appendix

MALDI-MS analysis

S. scincus – 17 kDa (upper band)

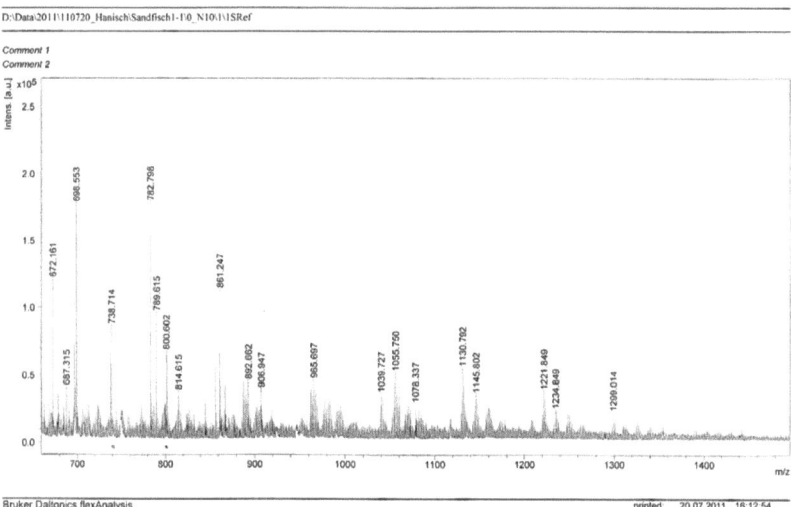

S. scincus – 17 kDa (lower band)

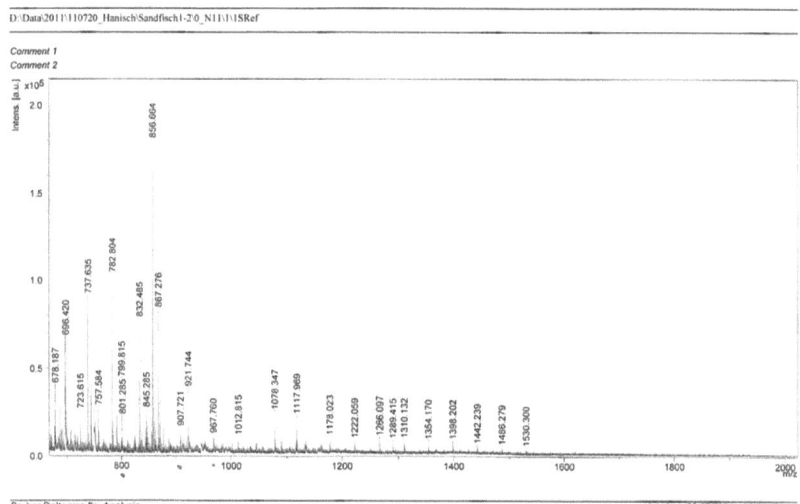

Appendix

E. schneideri – 17 kDa

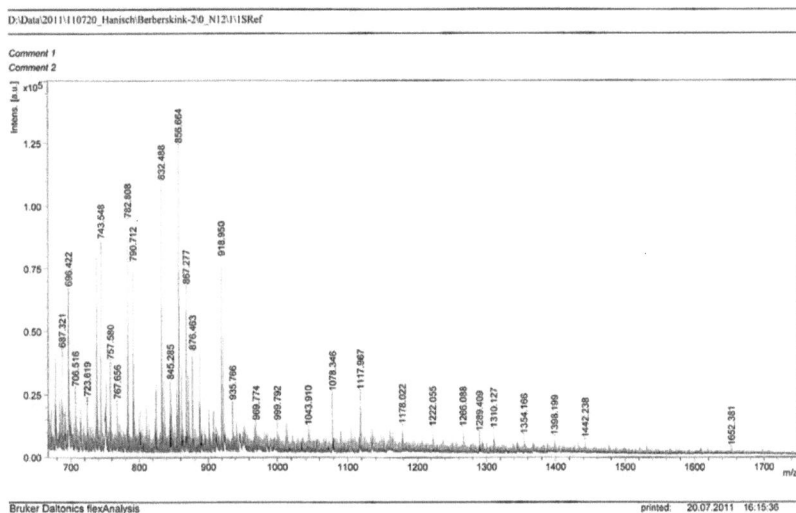

M. anchietae – 17 kDa

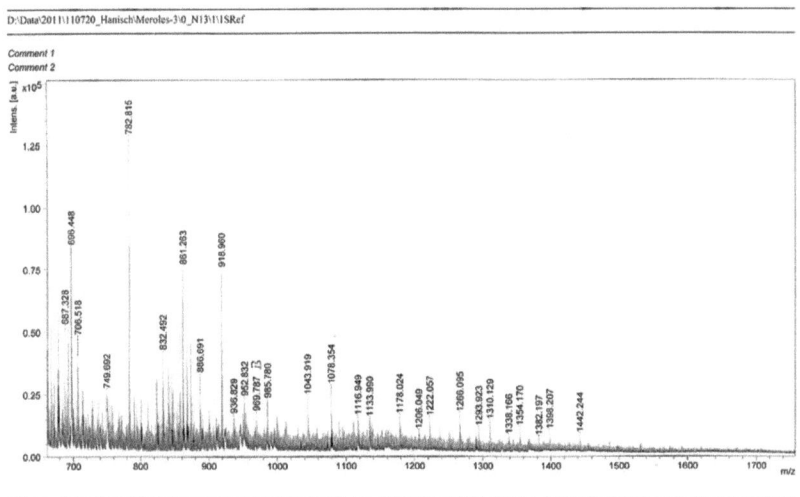

i want morebooks!

Buy your books fast and straightforward online - at one of world's fastest growing online book stores! Environmentally sound due to Print-on-Demand technologies.

Buy your books online at
www.get-morebooks.com

Kaufen Sie Ihre Bücher schnell und unkompliziert online – auf einer der am schnellsten wachsenden Buchhandelsplattformen weltweit! Dank Print-On-Demand umwelt- und ressourcenschonend produziert.

Bücher schneller online kaufen
www.morebooks.de

 VDM Verlagsservicegesellschaft mbH
Heinrich-Böcking-Str. 6-8　　Telefon: +49 681 3720 174　　info@vdm-vsg.de
D - 66121 Saarbrücken　　　Telefax: +49 681 3720 1749　　www.vdm-vsg.de

Printed by Books on Demand GmbH, Norderstedt / Germany